the architecture

of bridge design

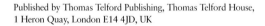

Published by Thomas Telford Publishing, Thomas Telford House,
1 Heron Quay, London E14 4JD, UK

First published 1997

Distributors for Thomas Telford books are
USA: American Society of Civil Engineers, Publications Sales
Department, 345 East 47th Street, New York, NY 10017-2398

Japan: Maruzen Co. Ltd, Book Department, 3–10 Nihonbashi 2-
chome, Chuo-ku, Tokyo 103

Australia: DA Books and Journals, 648 Whitehorse Road, Mitcham
3132, Victoria

A catalogue record for this book is available from the British Library

ISBN: 0 7277 2529 7

© David Bennett, the contributors and Thomas Telford Services Ltd 1997

Designed and typeset by Lawrence Kneath

Printed and bound in Great Britain by The Cromwell Press, Melksham,
Wiltshire.

the architecture

of bridge design

David Bennett

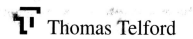

Thomas Telford

Contents

Introduction

by David Bennett

"The Architecture of Bridge Design" is a series of essays on bridges which attempts to articulate the process of conceptual design, the inspiration for an idea and how it has been developed into the final solution. It is not about analysis or structural calculation nor about torsion and deflection of a bridge. It is about exploring the functions of a bridge, defining purpose of place, putting people values in context, the spirit of creativity, impact on the environment, intuitive understanding, reasoned progression of an idea, exploiting material technology and construction innovation, tension between lightness and mass, and between sculpture and scale.

I chose the word architecture to mean the whole and not the parts, the design intent from concept to finished reality, and the contextual and structural clarity. The words of Mies Van de Rohe come to mind as a basis for good bridge design, where architectural expression is reduced to the essence of structural form, relying only on shape and scale to achieve elegance and beauty.

The notion that architecture and bridge design is one process is not new. The great age of bridge building at the turn of the century drew inspiration from classical architecture in much the same way as for buildings. The medieval and roman bridges that remain standing today are admired as much for their artistry and monumental form, as for the superhuman effort required to build them with such limited technology. The arrangement and construction of these bridges respected a certain rigor and order in the ornamentation of design; it was part social culture and part tradition.

The effort and human endurance needed to build these structures, have coloured our perception to such an extent that we believe them all to be masterpieces. A few of them were masterpieces most definitely, but many were hugely egotistical or hurriedly built for expedience and not for aesthetic pleasure.

But what is good design today? To those who have only seen the drab and humdrum offerings up and down the M1 and M4 corridor, it may be the celebrated bridges of the past seen in well illustrated books on bridges, like the works of Telford, Brunel, Roebling, Eiffel, and Amman, or the Pont du Gard built by the Romans. For those who have travelled extensively in France and Switzerland and are aware of modern bridge design, good design might suggest the work of Jean Muller, Eugene Freysinnet, Robert Maillart and Christian Menn. In Germany it could be Jörg Schlaich and Fritz Leonhardt, and in Spain it might be the work of Calatrava and Cassado.

Each period in history will no doubt uncover monsters and marvels of bridge engineering, as they do with buildings, so that succeeding generations can learn to distinguish between good and bad design. In looking at the past, we should not make the mistake of copying; instead we should be improving what has been built. Today we see many bridge designs that are stilted shadows of an overplayed bridge form, tried and tested many years ago but not intelligently remodelled. Like coffins with an empty promise, they straddle our motorways, rivers and canals in faceless anonymity. These humourless structures have tempered condemnation by reasons of economy, ease of construction and conforming to a set of inflexible rules; rather than proportion, scale and character.

It simply is not true that an elegant bridge is more expensive to build than a plain "we've done this one before" design. Robert Maillart and Eugene Freysinnet, both pioneers of modern concrete bridge building technology, won bridge design awards because their bridges were more economical to build than those of their competitors. They created some of the most elegant and charismatic bridges seen in the early 1900s — the Salginatobel near Schiers in Switzerland and Plougastel Bridge in Brittany over the Elorn Estuary. Each

understood the place of architecture in design, the ability of concrete to be moulded into the most economical structural shape, and the control of the dynamic forces that act on a bridge.

The bridges in this book reaffirm the basis for good architecture in the design process, in the quality of the space a bridge occupies and its linkage and connections with the environment.

The arguments for good design are presented through the personal accounts of architects and engineers responsible for award winning bridge ideas, accompanied by preliminary sketches, scheme drawings and computer enhanced images.

Now at last we are seeing talented architects being encouraged to collaborate with bridge engineers, to breathe new life into bridge design, to create a better vision for

tomorrow. In France that vision has existed for a number of years, as the examples in the book will illustrate.

In England this creativity had its genesis in the early 1990s with two significant, if controversial events; the Bloomers Hole Bridge Competition which was won by Cezary Bednarski in 1992 and a design idea for the East London River Crossing by Santiago Calatrava in 1990. They were both exciting and imaginative ideas, if a little irreverent perhaps. They still remain unbuilt and without patronage, because of the controversy that surrounded them at the time, although they are very much admired and talked about today.

The finality of good design can still seem an arbitrary decision and one of personal preference, a rationalisation of one's inner thoughts and impressions, based on past

experience. Like music, we may hear the same notes, but we interpret them with different emotion.

The judgement of good design is still a subjective one, but it is hoped that as the observer casts a critical eye over a design, an attempt is made to understand the design intent in order to give a more considered response.

I have used extracts from Herbert Reed's essays on "The Meaning of Art" and Edward de Bono's ideas on lateral thinking in "Six Thinking Hats", as opening statements to some of the bridge subjects. They have helped me to express the emotion that I experience when seeing something of beauty for the first time.

David Bennett, 22 August, 1996.

Cesary Bednarski's controversial pencil-thin, carbon fibre bridge for the Bloomers Hole Bridge Competition.
Lord St John of Fawsley, The Royal Fine Art Commission and the judging panel described it as a "beautiful solution of great simplicity and elegance entirely appropriate to its rural setting".
The residents of Lechlade labelled the design "a yuppy tennis racket from hell".

Good Design is not just
Engineering

by Cezary Bednarski

Model of the Bloomers Hole Bridge.

Engineering per se is not a pure, definite science in the same way that astronomy and mathematics are. There are many answers to an engineering question — materials can work in tension or in compression or a combination of both and still perform the same function. It is the designer's intuitive sensitivity that limits the number of engineering options at his disposal, before refining them into solutions that turn engineering into an exact science. Occasionally the creative process demands new images or idioms, thus pushing out the frontiers of engineering technology.

Designers endowed with such sensitivity reach depths of reasoning and meaning. The structures are timelessly rooted in their location and designed to the technical requirements and aspirations of both the client body and contemporary society. Such designers realise the emotional and social power of their work. Inevitably their structures acknowledge that the world, whether natural or man-made has been there before them — their designs exude connections and an equilibrium with the past. This sensitivity is by no means exclusive to any profession, although it has to be recognised that architects, more than engineers, are better able to cultivate this gift because of their training in architecture.

That said, it is regrettable that a growing number of contemporary bridges and buildings are too simplistic in design; they are 'one-liners' devoid of any sensibility or meaning, other than the materialism of society. They are the products of technocrats and administrators preoccupied with the rationale of structural integrity, functional and economical rectitude. They are devoid of any real reference to the human spirit. Such structures impoverish our environment and our future heritage.

What is Good Bridge Design?
Bridges You Can Touch

by Jörg Schlaich

Nothing restricts imagination more than rules and regulations which state "It has to be done like this." Of course when you are designing and building a pedestrian bridge for instance you have to know and to use the experience gained from designing road bridges and rail bridges. You should then detach yourself from them, so that you are not restricted to the few forms that have become generally accepted in the field. Restrictions that are imposed by fast traffic and the economic constraints imposed by a materialistic society, which do tend to kill creativity.

Big bridges for traffic, are experienced from a distance like a sunset, but pedestrian bridges are confronted head on — you can walk over them and you can touch them. They are part of the furniture of a city, and sculpture in an unspoilt landscape.

So pedestrian bridges must be on a human scale, slender and filigree, but built just as durably and as easy to maintain as big bridges.

In functional terms a shallow deck is sufficient for the

Motorway footbridge, Sindelfingen 1986.

walkway of a footbridge. Everything else — by that I mean the support structure — is a means to an end. But the support structure need not be enclosed or solid, it could be transparent. It could be broken down into poles, ropes and masts in such a way that the deck defines the scale of the bridge.

Why should arched bridges be used only for big spans of around 200m, or cable stayed for spans of 400m, or suspension bridge spans exceeding 800m? Is diversity and variety not admissible in bridge design?

Uniformity and repetition leads to boredom of expression in bridge design. A meadow full of wild flowers is surely more beautiful than a bed of tulips, however colourful they might be. Should the least building cost be the only criterion for designing a bridge? Should we not look at the project as a whole, from the use of land and resources, through to the integration into the landscape, the demolition and even the recycling of a bridge?

Is bridge building not part of our building culture?

transportation**bridges**

part *one*

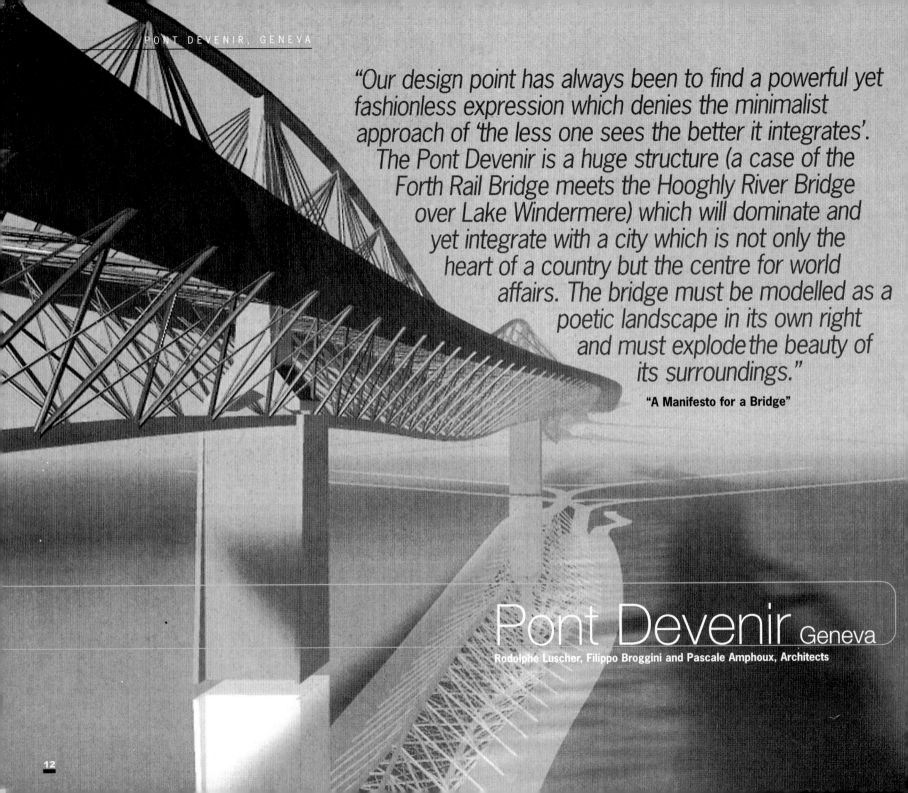

"*Our design point has always been to find a powerful yet fashionless expression which denies the minimalist approach of 'the less one sees the better it integrates'. The Pont Devenir is a huge structure (a case of the Forth Rail Bridge meets the Hooghly River Bridge over Lake Windermere) which will dominate and yet integrate with a city which is not only the heart of a country but the centre for world affairs. The bridge must be modelled as a poetic landscape in its own right and must explode the beauty of its surroundings.*"

"A Manifesto for a Bridge"

Pont Devenir Geneva

Rodolphe Luscher, Filippo Broggini and Pascale Amphoux, Architects

The crossing of "La Rade" at the southern end of Lake Geneva, is a solution tailored to solve the traffic problems of Geneva. It is an idea that explores the multiplicity of uses for bridges. And yet it is more than a bridge and more than a crossing, it is an island city and the 21st century "Florentine Ponte Vecchio". It captures a moment in the event of making the crossing by motor car, by walking, seated inside a restaurant, strolling through a shop, or just watching the events.

The combination of uses of such a bridge demands the design development of new and daring technologies and a thorough examination of the construction possibilities. The structure we have conceived is on a vast scale because it must connect one bank to another, at the most efficient point for decongesting the traffic flow on either side. If it does not solve that problem it is as dead functionally as the dodo. Let's put aside the dilemma of what the best connecting structure should be … a tunnel or a bridge? Let's explore the poetry of bridge design first.

Theme and Variation

A bridge structure of this size, with four spans each of 300m, must integrate with the spirit of the city which is not just the heart of a country, nor the meeting place of a continent, but arguably the centre of the Western world. This bridge must be modelled as a poetic escape from the tensions of the city itself. At one and the same time it

Left. Where the lake becomes the Rhône a plethora of crossings, the one meant to alleviate the congestion of the next, has resulted in a charming strangulation of the existing transport system. Like the water jet, where air is needed to oxygenate the larger waters, the breathing structure of the Pont Devenir could rescue a suffocating city.

Below. Anchored to the left bank, the structure is pulled like an extended spring into the underworld of the urban park on the right bank—a total experience charged with differential views, movements, sensations and sounds.

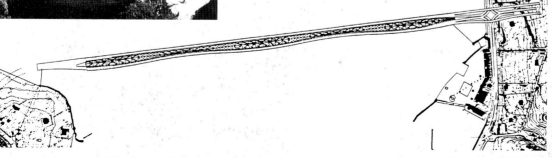

reaches for the sky, it dives into the lake and by its very form it must literally explode the beauty of its surroundings.

The bridge is located between opposite banks of Lake Geneva; banks which are obliquely symmetrical to one another. The relationship between water and city deviate differentially either side of the crossing. The right bank is fringed with greenery, with an urban park of subtle colour and fragrances—an ecological island within the city which must remain intact.

So as not to spoil this side, the bridge disappears into a tunnel before it reaches the bank, diving under the parkland fringe. The left bank on the other hand is a dense, built up environment of buildings and motorways, overshadowed by a muscular mountain backdrop. On this side the bridge anchors itself to the mountain, to link with the embankment by a crossed interchange.

Playing on the notion of co-existence with differential speeds of travel, the structure incites the mixed modes of travel, as well as paths of travel. Running on the carriageway only located on the outside fans of the bridge, the cars leave the safe inner zone to the pedestrians. The footpaths in, above and under the structure allow a variety of routes and walkways away from the noise of the traffic.

The footpaths, with their transparency or opaqueness, their shining or matt surfaces, their open or enclosed spaces, offer enjoyment of unexpected views and perspectives of the lake, the shoreline, the city and the mountains. By limiting the travelling speed of vehicles, the bridge makes crossing "La Rade" by bus an attraction in itself. One could even imagine a bridge transport system, a mono-rail or cable car system integrated into the structure, with stops every 300m

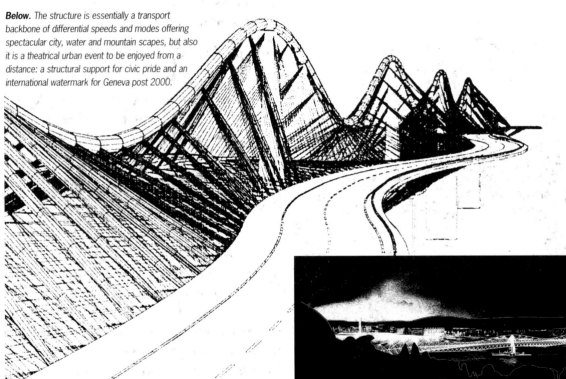

Below. *The structure is essentially a transport backbone of differential speeds and modes offering spectacular city, water and mountain scapes, but also it is a theatrical urban event to be enjoyed from a distance: a structural support for civic pride and an international watermark for Geneva post 2000.*

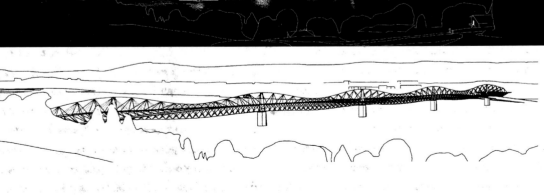

connecting into the pedestrian routes through the bridge.

By virtue of the spaces created within and surrounding this multi-faceted structure, which is basically a continuous girder frame of variable inertia, the bridge also becomes the support for an infinite number of different functions, be they permanent or ephemeral extensions of the city.

Form and Function

With traditional construction, form follows function. A conventionally built bridge is no exception. But here we are not talking about a conventional idea for a bridge. The purpose of the bridge is not just for taking traffic across from one side of the city to the other without getting wet. A tunnel or immersed tube structure could do that just as well but it would cost three or four times more and constitute a sunken and invisible investment to be enjoyed only by the motorist. But even for the motorist there is no visual beauty to relieve the dullness of the enclosing walls during the subterranean journey. So it does make economic sense to build a bridge. Such a bridge should also embrace the city, it should provide a place for people to stop and shop, or meet and eat.

This combination of uses demands the development of new and daring technologies. The structure of the bridge is a continuous girder of variable inertia interconnected by mechanical gusset plates, which allow for linear and spatial variation of structural forms. The triangulated configurations of the structure vary according to their function. Over the pier supports the structure deepens to become a series of five sided frames tied by tubular steel sections and braced across the middle by a steelwork lattice. To carry the outer roadways a series of trusses spring from the central triangulated frame like

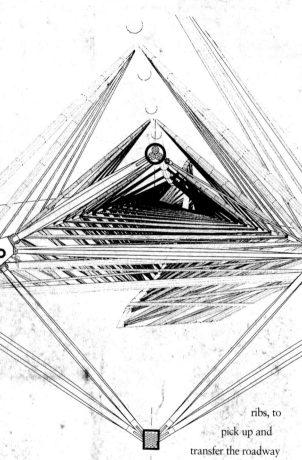

Below. *Two independent carriageways on opposite sides of the bridge, offering two very specific travelling experiences. One an undulating drive, the other a meandering crossing.*

ribs, to pick up and transfer the roadway forces back to the main structure. A tubular steel duct full of high tensile cable strands runs continuously over the top and bottom line of the bridge frame to transfer the load of the span to the masts over the pier supports. In essence it is a composite cable suspension bridge of quite stupendous proportions and structural igenuity.

The geometry of the principal support structure allows for the variation in attitude of the harnessing points as well as

their position in plan. This offers the possibility of running two independent carriageways on opposite sides of the bridge. The structural elements in tension are the top cable strand, the bracing lattice frame and the lower cable duct. The elements under compression are the triangulated elements of the central frame. The resulting structure gives the spatial characteristics we were looking for as well as the strength to span 325m between supports.

Reality and Utopia

If anyone thinks that this structure can be built at great speed and for very little cost then think on. Four decades of hesitation, of doing nothing, on the part of the Geneva authorities in order not to respond to what is after all a desperately needed amenity has reached a real impasse in a beautiful city. So what are we saving and what are we sacrificing in the name of economy by doing nothing, I ask myself!

Geneva, the host city to numerous world organisations, to world leaders, international companies, the rich and the famous, the tourist and the sightseer, should be able offer her local clientele and her esteemed guests a potentially exhilarating and challenging concept which can satisfy the cultural, gastronomic and touristic needs of the day. It will also resolve a major transport problem.

So then if the political willingness would only crystallise, this long wait almost into the next century will be the creative moment of a real groundbreaking event.

"The foundation of reality is Utopia. Utopia is always in advance of reality. So the 'Pont Devenir' awaits Geneva. How much longer will Geneva let her pass by?" Rodolphe Luscher.

Kirchheim Motorway Overbridge

Jörg Schlaich, Engineer

The quality of a structure depends quite fundamentally on the structural design of its details. Force flow at geometrical and statistical singularities must be pursued carefully and translated into material in such a way that tension peaks and abrupt deviations are avoided.

The Visible Clues

Bridges over roads and motorways offer an outstanding opportunity to show large numbers of people what good bridge design looks like. Unfortunately the opportunity presented to the engineer has not been fully exploited. It is true that nothing restricts imaginative design more than rules and prescriptive statements on how bridges should be built. This stems from an economic doctrine on bridge design imposed by highway authorities, whose remit is to keep expenditure down. This has led unfortunately to designs lacking in creativity, lacking in visual clarity. It is a mistaken belief that such designs represent value for money.

The many examples of pedestrian bridge design in recent years could serve as a model for creative bridge design on a larger scale. This is how I approached the design of the motorway overbridge for Kirchheim. The new bridge was needed when the existing autobahn was widened to six lanes. The autobahn lies in quite a deep cutting which the bridge was to span. There was enough room to design an overbridge whose support structure would be visible under the span. The design that I conceived was meant to be taken in quickly, as the motorist passsed by at high speed. Therefore the structural form had to be understood at a glance, and the flow of forces emphasised with visible clues.

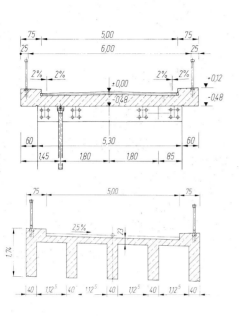

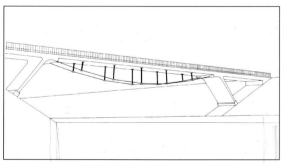

Structural Form

The familiar bulk of a heavy looking box girder bridge riding on inclined piers—so common on many European motorways—can be made to look slim and interesting by making the structural assembly work harder. I thought it would be interesting to underpin the bridge deck with an arrangement of tension cables and struts holding up the span.

This can be a very durable and robust solution for a bridge. The slender structure supporting the heavy bridge deck was visually expressive and its logic instantly recognisable. But unfortunately the transparency of the design was too avant garde for the highways authority. The overriding concern was about vandalism and the exposure and vulnerability of the deck support structure to potential sabotage! So the tension cables and struts were encased in

concrete like most prestressed concrete structures.

However I did manage to retain the profile of the original design, although the curvature of the span was not as pronounced over the mid span section. In order to reinforce the plastic lines of the bridge, the surface of the structure was given a very smooth concrete finish. This also meant that, quite deliberately, a multiple web T-girder section was formed rather than a box girder.

Pont Isère *the Tulip in Bud*

Alain Spielmann, Architect

"On a clear sky blue day, the view of snow capped mountains seen through the changing moiré patterns of the cable fans presented the greatest optical art show on the planet. Victor Vasarelli & Bridget Riley eat your hearts out! The tall central pylon captures the attention immediately, becoming the reference point for everything around you, channelling the eye along the bridge, the cable fan and back. The pylon appears to thrust out of the earth like an arm of Hercules with fist clenched, as though holding up the cables to prevent the bridge from crashing into the valley below. Seen from another angle the pylon resembles a tulip in bud. *The contrasting imagery is beguiling.*"

First impressions of Isère Bridge near Romans, on the A49 trunk road to Grenoble. David Bennett, May 1994.

Photo: Grant Smith

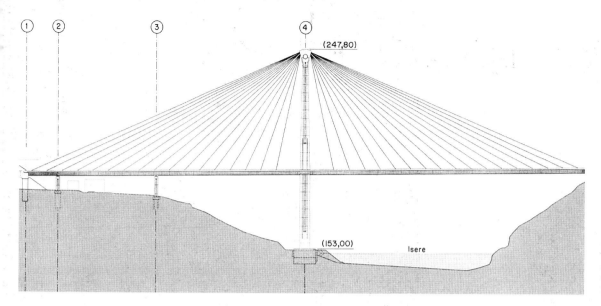

① ② ③ ④ (247,80)

(153,00)   Isere

Designed to sew back the two banks of the Isère Valley, Pont Isère carries the A49 across the 300m span.

The Two Halves

When I looked at the site of the bridge in this beautiful part of France, it was clear to me that an arch or flat box girder bridge would not have the grace or continuity of a cable stayed structure. The scenery here was majestic. The bridge was an important link to the two banks. It must join the two halves as though it was meant to be there in the first place. To create the feeling of lightness and transparency of the structure, the pylon shape and proportions were the central core of the design. One unifying piece was what I had in mind. In the past the disjointed profile of many cable stayed bridge pylons in France have looked odd. They appear slim above the bridge deck yet very squat below it.

The new bridge must straddle some 300m across opposite banks of the Isère River to link the A49 trunk road to the Valence motorway in the Rhône Valley and to the A48 which runs into Grenoble. The crossing takes place near Baume d'Hostun and Saint-Lattier at a point where the river bed is 80m wide. On the east side the bank slopes gently towards the river, reaching it almost horizontally for a short way. There are a few houses, some agricultural buildings, a local road and a railway line cut into the moderate slope that runs parallel with the river.

On the opposite side, the bank is steep and heavily wooded almost down to the water's edge. It would be easy to build the support piers on the shallow east bank, but any construction on the west bank would be problematical. The engineers looked at several approaches to positioning the support piers. Two alternative ideas summed up the difficulty of the situation.

One solution envisaged the construction of the piers on both sides of the river. This created a structure with a main span of 100m and the need for an access road on the steep west bank. The bank did not appear stable enough to build an access road. It would be very costly to construct and would spoil the natural look of the river bank by rooting out the trees and scarifying the landscape. The other option considered the possibility of building the pier on the shallow east bank, near the edge of the river, and setting down the next pier support on top of the steep bank, about 150m way. This eliminated the problem of building on the steep bank.

A geotechnical investigation of the site showed that the river had cut a deep channel into the bedrock and as a result the bedrock was now overlain with deposits of gravel. The deposits varied in thickness from 13m on the east bank to 4m on the west bank. If the pier support was built on the east bank it would only require a raft foundation 4m deep and 11m in diameter, contained by a shallow cofferdam. For the steep-sided bank the abutment and intermediate piers would have to be piled to a depth of 10m or more.

The Eye of the Needle

The idea of a bridge with a 150m span and central pylon was a logical solution. The central pylon set on the edge of the east bank divides the structure roughly in half, being 164m and

Construction of the box girder deck showing the stiffening septums.

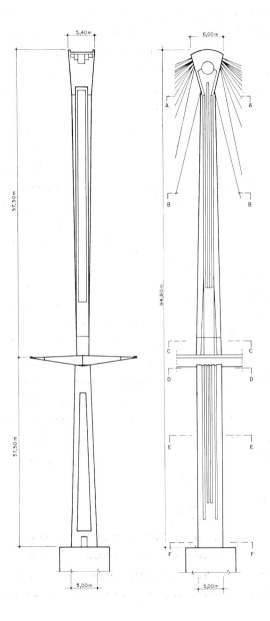

140m respectively between east and west abutments. The proportions of the span to height create the potential for an aesthetically pleasing composition.

These options came about after very fruitful discussions with the engineer, Jean Muller, who had just finished his beautiful Sunshine Skyway Bridge in Florida and was convinced of the beauty of central stays. The question arose: what kind of solution should be used for the anchorage of the stays — harp or fan stayed arrangement? The architect proposed to study a solution not recently used in France, inspired by the Pasko Kennewick American bridge, with a saddle on top of the pylon.

A bridge with only one continuous pillar in the axis of the deck, a saddle on top, without discontinuity between the two parts under and above the deck, were the features adopted together. The hole in the top of the mast was a solution to express architecturally the fact that the stays were passing from one side to the other without interruption.

The conception of the deck, with its triangular shape, is obviously the signature of Jean Muller.

Since the deck is suspended by stays along the centreline, the cables were anchored to the septum of the deck structure at 4m intervals. The box girder deck is made up of a top slab of 21.4m — the width of the carriageway — and a central spine 2.4m deep, and an angled bottom slab creating a triangular cross section. The edge and inner spine sections of the deck form a true box structure with the top and bottom slab, which combine to generate great torsional stiffness. The triangular profile of the deck also improves the aerodynamic resistance of the deck structure, reducing the drag coefficient. Curved black metal cowls over the extreme edges of the deck hide the galvanised metal rainwater gutters and emphasise the leading edge of the aerofoil.

Without a shadow of doubt the central pylon, the profiled features to the shaft, the hole at the top and the saddle for the cable stays, were going to be the real challenge for both the aesthetic and engineering efficiency of the design. To create the unity of the structure and to maintain its lofty presence, the longitudinal and transverse profiles have been detailed quite differently. In essence the central pylon is a cast in place hollow concrete shaft 94.8m high, 57.3m of which rises above the bridge deck. Seen sideways on, the pylon tapers gradually along its length until it meets the flared head of the saddle where it is then punctured with a hole, like the eye of a needle. The scalloped features to the pylon face create shadowlines that heighten the grace and slenderness of the structure as the eye is drawn to the saddle head.

Isometric section and cut away showing the principle prestressing cables of the deck structure.

"In the distance is the soft green silhouette of the Alps, in the foreground the lush green fields of farmland and fruit trees carpeting the countryside for miles. The dual carriageway ahead looks a little strange. Down the central reservation is a mast rising like a menacing cobra about to strike. We are past it in a flash. A moment of wonder is played back to the mind as we crossed the bridge and a gorge opened up to reveal the face of a silvery river below. Over the bridge the carriageway takes a sharp right. We double back to the next junction to find a way of getting a closer look at the bridge. Zigzagging the car down country lanes, the occasional glimpse of the cable fans, peeking through gaps in

the trees, spurring us on. Suddenly the whole bridge comes into view … it takes your breath away."

In the transverse direction, seen from the centreline of the carriageway, the pylon is a mirror image of itself above and below the deck. Wide at the base and wide at the top, it curves gently inwards as it narrows in width to reach the bridge. The edges are profiled once again to sharpen the contrast, emphasising the symmetry of the shape. The slightly shorter length of the pylon section sits below the deck, to reassure the onlooker that it is positioned the right way up!

The pylon head itself was built in two sections. The first half was cast to the base of the head containing the saddles for the stays. The top section encloses the saddles and the steel frame that supports the saddles. Precast fascia panels were placed within the formwork after the reinforcement cages were positioned and before each section of the pylon head was cast. The saddles were clustered in groups of differing curvature from 2.5m, 3.5m to 5.0m in radius corresponding to the alignment and displacement of the cables. The saddle profiles were made of precast concrete and fixed to a steel truss that was bolted to the lower pylon head.

The Stays

The fan configuration of the stays is created by converging the cables over the pylon head, each cable directed by the saddle pathways that fix their trajectory and angle of inclination to the bridge deck. The advantages of a fan arrangement of cables over

The pylon head showing the precast concrete saddle guiding the cable pathways.

All photos: Grant Smith

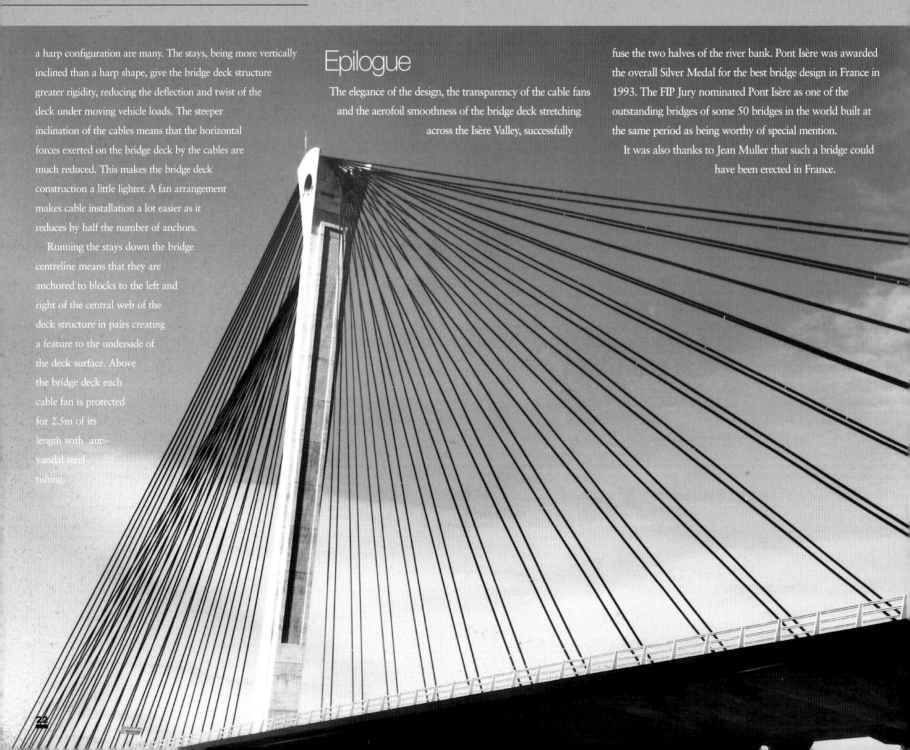

a harp configuration are many. The stays, being more vertically inclined than a harp shape, give the bridge deck structure greater rigidity, reducing the deflection and twist of the deck under moving vehicle loads. The steeper inclination of the cables means that the horizontal forces exerted on the bridge deck by the cables are much reduced. This makes the bridge deck construction a little lighter. A fan arrangement makes cable installation a lot easier as it reduces by half the number of anchors.

Running the stays down the bridge centreline means that they are anchored to blocks to the left and right of the central web of the deck structure in pairs creating a feature to the underside of the deck surface. Above the bridge deck each cable fan is protected for 2.5m of its length with anti-vandal steel tubing.

Epilogue

The elegance of the design, the transparency of the cable fans and the aerofoil smoothness of the bridge deck stretching across the Isère Valley, successfully

fuse the two halves of the river bank. Pont Isère was awarded the overall Silver Medal for the best bridge design in France in 1993. The FIP Jury nominated Pont Isère as one of the outstanding bridges of some 50 bridges in the world built at the same period as being worthy of special mention.

It was also thanks to Jean Muller that such a bridge could have been erected in France.

A pedestrian's view of the bridge from the west abutment.
Photo: Grant Smith

A75 Clermont Ferrand Viaduct

Alain Spielmann, Architect

Photo: Grant Smith

"A design is not there to seduce or fascinate egotistically, it must seek the truth in the aesthetic of design and to capture the heart of its intrinsic beauty." Thoughts on the architecture of bridge design of Alain Spielmann could so easily have been spoken by Gustav Eiffel some 100 years ago when looking at the site of his famous Garabit Bridge, which strides majestically across the River Truyère, a thousand metres downstream from the A75 crossing, the site of the Second Garabit Bridge.

In the Wake of Eiffel

The challenge of building a modern motorway bridge in full view of Eiffel's Garabit Bridge, trying to do visual battle with such a beautiful wrought iron arch, needed thoughtful and careful design consideration. The conceptual design must conform with the economy of modern motorway construction, but at the same time this bridge had to relate to the historical setting of Eiffel's design and to the scale and topography of the magnificent Truyère Valley below.

For me this bridge, amongst the many bridges I have designed, was going to represent a statement of an age, of how we have progressed as bridge builders in 100 years and as creative designers. How do we make our choice of structure? How do we adapt the structure to the site and the area? What are the parameters to guide us to the right decision?

In many ways the trial and error approach of scheming alternatives for their ease of construction and design economy is a good starting point. Three preliminary scheme designs were made — a three span box girder bridge deck supported on vertical piers; an arched bridge with the bridge deck running on the top, a bit like Eiffel's Garabit; and a box girder bridge supported on inclined piers. A cable stayed option was eliminated before the preliminary design review as it would look totally out of place and out of context in this setting.

The economic choice was naturally the box girder bridge supported on vertical piers, followed by the option with inclined piers. An arched bridge proved to be the most expensive. Now we had to decide on the aesthetic quality of the overall design and how we could enhance the composition to make it sit well some 80m above the river, in full view of Eiffel's Garabit Bridge.

The valley to be crossed measured 350m from opposite

Above. Eiffel's celebrated Garabit rail bridge, just 1000m downstream of the A75 road bridge.
Photo: Grant Smith

The preliminary design options superimposed over the River Truyère.

Below right. Box girder with vertical supports.

Below. Box girder with inclined supports.

Bottom right. Arched bridge scheme.

abutments, on a plateau of the Central Massif. The river banks are steeply sloping, at an average angle of 45 degrees. The width and depth of the river did not matter much because the river is crossed 75m above the high water mark and flow is regulated nowadays by the Grandval Dam upstream.

We were mindful that the bridge would come into service on the hundredth anniversary of the Eiffel Bridge. A harmonious integration with the site and with the Eiffel Bridge was paramount. We suggested to our client the creation of a car park area and footpaths around the new crossing, from which to view both bridges. In summer there are considerable numbers of tourists in the area, many taking a lazy boat journey along the Truyère, to enjoy the rock formation, steep gorges, lush landscape and the novelty of Eiffel's Garabit Bridge. The quality of the bridge finish and construction details was going to be critical for the new bridge structure!

Some Snow & Sticks

The three bridge solutions that we had chosen were superimposed on a photo montage of the crossing to make a preliminary judgement on the aesthetics. The arch bridge was immediately rejected. It appeared to mimic Eiffel's bridge and to compete with it for scale. The three span concrete box girder with inclined piers was chosen because it looked better than the stilted arrangement with vertical piers. The client was insistent that an aesthetically pleasing scheme was to be preferred. Cost was not a design criterion.

Up to this point the engineers had driven the design. As the bridge architect we took over when the functional form of the bridge structure had been agreed. We had to work pretty hard to reduce the squatness of the engineered solution. It was nearly a year before the

The construction sequence

(a) the temporary support shaft

(b) foundation, falsework and decking for the inclined pier

(c) balanced cantilever construction of one half of the bridge deck

(d) completing the other half of the bridge deck

(e) the finished bridge after demolition of the temporary shafts.

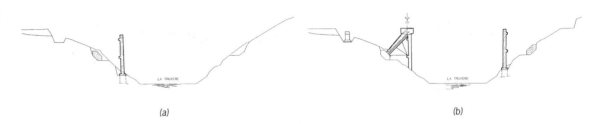

(a)   (b)

scheme drawings were actually finalised for tender submission.

Solid piers on a sloping bank with a fairly level bridge deck did not offer much visual drama in my opinion, no matter how well they were detailed. Such piers did not create a balanced horizon for the span. Splayed support legs were the preferred solution, but they would be more difficult to construct. Having read about a new bridge being built in Brittany which had splayed legs, I thought I would make a site visit to study the construction problems.

With the firm that erected the bridge, I noticed how difficult it was to build the inclined supports. I spoke to the contractors to get a better feel for the problem and how these difficulties could be avoided if they were to build them again. We modelled the splay of the piers to improve the harmony of the structure and to add interest

to the shape and profile of the piers. The two legs may appear simple in shape, but it took a lot of complex designing to make it so. The pier supports were tapered and splayed to present intersection parallels with the deck. The modelling and profiling of the pier legs were sculpted to create the drama and tension of a ski jumper just before take off.

Seen from the longitudinal direction the two piers get closer as they climb to meet the bridge deck. Seen side on, the pier grows progressively wider as it nears the top. Both sections relate to the bridge span and the scale of the river valley in quite different ways. In the side-on view we see a slim leaping structure with a strong joint, connecting the bridge deck to the pier. As it tapers down to the base it helps to complete an arch with the central deck, creating an added line of interest. Looking at the bridge from the other direction the piers are wider at the

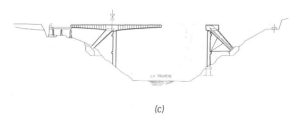

(c)

(d)

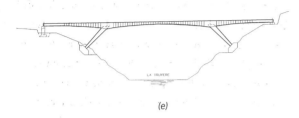

(e)

base, like the trunk of a tree. This was done to lighten the appearance of the structure and to highlight the curve of the bridge deck section.

We chose a green aggregate which gives the concrete a greenish hue. We detailed every panel of formwork so that the joint patterns would appear well ordered when the leg was cast. Greenish precast panels were recessed in the pier legs to emphasise the taper of the piers and to split them into three parts to create thinner proportions. Rectangular features recessed near the tops of the piers indicate where the temporary supports were fixed during construction of the inclined piers.

It is worth noting that the shape and form of the bridge that was built could have been constructed in steel. We worked with two engineering teams to develop tender documents for a steel and a concrete design. We thought the steel option would prove to be the cheapest because it

would be lighter and easier to launch across the valley. However the cost of long term maintenance was a big disadvantage for the steel scheme and in the end the better life cycle cost of the concrete option proved decisive.

The Construction Sequence

First the foundations for the temporary support shafts were cast on each bank. Then the temporary support structures were built. Each measured 4m by 4.5m in plan comprising a reinforced concrete shaft 45m high, which was hollow in the centre. The temporary piers were necessary to support the main pier scaffold during construction and to support the bridge deck. After the inclined piers on the west bank were constructed, the balanced cantilever construction for the box girder bridge

then commenced for one half of the bridge shored up by the temporary support shaft. The second pair of inclined piers was then built on the opposite bank supported by the temporary shaft, with the other half of the bridge deck construction following on.

The bridge deck structure is a prestressed concrete box girder with an extended top section 20.5m wide, which forms the dual carriageway. The box girder is supported by two webs 600mm thick and inclined at 30 degrees. The inclined webs arch between the pier supports reducing in depth from 11m at the supports to 3m at mid span. The bottom slab of the box section varies in width from 12m at mid span, down to 8m over the supports. The visually contrasting profiles created by the changing deck section give the structure its momentum, like a skier in flight—shallow and wide at mid span, deep and narrow over the supports. When the two halves of the

Above. *The temporary shaft being demolished.*

Right. *The splayed piers of the bridge connecting as one segment to the box girder bridge beam. Photo: Grant Smith*

bridge deck were connected, the temporary support shafts on each side of the bank were demolished. The full force of the bridge had to be transferred to the inclined piers before demolition, by removing the hydraulic jacks that propped the bridge deck off the temporary shaft head.

On July 3 1993 the bridge was opened by President Mitterand and marked the completion of the A75 motorway nicknamed "La Méridienne" in the Auvergne region which links Clermont Ferrand and Beziers. It has been acclaimed as an exceptional engineering construction and complements, but does not compete with, the famous rail bridge of Gustav Eiffel at Garabit.

It is a fine statement of the art of modern bridge design. In 1993 the A75 Clermont Ferrand Viaduct was awarded the Ministry of Transport Gold Medal Trophy for the best bridge structure in France, ahead of Pont Isère which was placed second.

The Pylon and the Golden Section
The Counterpoint of Shifting Scale
Ronald Yee, Architect

"Good design has an imaginary point of reference, very like the centre of gravity, and around this point the lines, surfaces and masses are distributed in such a way that they rest in perfect equilibrium. The structural aim of all these modes is harmony, and harmony is the satisfaction of our sense of beauty."

Herbert Read, 1951.

In recent years there has been a revival of the old debate between architects and engineers about the appearance of structures. The aesthetics of bridges in particular. It is not sufficient that designs are just functional and structurally correct, nor is it acceptable for the advantages of modern technology to be exploited for economic benefit without careful consideration of appearance and aesthetics.

When the engineer has exhausted himself in minimising a structure quantitatively he cannot expect his science and technology to be a guarantee that his minimalism will result in beauty which will be appreciated by all. Similarly when an architect has sculpted a structure he must not be surprised that his art will not be appreciated by the economist. Clearly, beauty resides in the mind of the observer and appeals to particular emotions.

Beauty is the quality which gives pleasure and joy. No true engineer or architect would deliberately set out to create an ugly structure. Can such intangible characteristics be described in terms that have universal understanding? This is what we strive for when we address a design concept. It is not the only way a designer can develop an aesthetic appreciation. Intuitive understanding of harmony and of proportion, the gift of design sensitivity, of application and learning as well as a talent for perceptive observation, are the other essentials for good design.

Beauty and Proportion

Aesthetic appreciation and visual beauty are difficult concepts to define. Aesthetics or the science of perception is concerned with the material qualities we perceive—colour, texture, tone, smoothness and so on, and the physical reaction to what we see in the arrangement of the shape and form of the materials.

Certain arrangements in the proportion of shape and form result in pleasurable sensations. Lack of it leads to a reaction of indifference, discomfort or even revulsion. It is possible that some people are quite unaware of proportions in the physical aspects of things. Just as some people are colour blind, so some may be oblivious to the sensory stimulus of shape and

bridge design or building architecture the different interpretations of beauty over the decades has engendered fluctuating and contradictory meanings. Surely it would be useful to find an aesthetic standard or set of values that provides useful criteria in design, which result in a consistent visual vocabulary.

The concept of beauty chosen in this study was formulated in ancient Greece and was the offspring of a particular philosophy of life. That philosophy exalted all human values, saw in the gods nothing more than man made larger than life. Art as

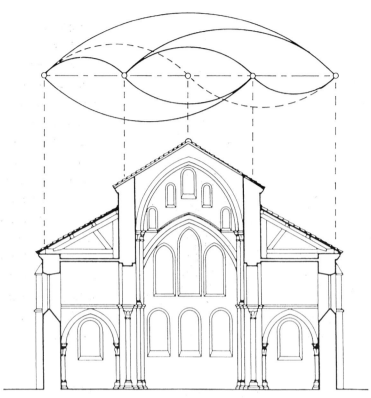

"Beauty is the quality which gives pleasure and joy"

form. It is probably more reasonable to asume that people unaware of aesthetics are rare. It is more likely that they have an underdeveloped perception of aesthetics.

Can we arrive at a universal law that can nurture in any one of us, no matter how poorly developed our sensibilities, a better appreciation of beauty? Can we somehow eliminate subjective judgement between people? Are we able to agree what is beautiful and what is ugly in design?

Aesthetic beauty may have dozens of interpretations and meanings, but the keystone is the one that can be measured and interpreted in a rational way. As we study the art of

well as religion was an idealisation of nature and man was the central point in the process of nature. This was the classical art of the Aphrodite of Melos and the statue of Zeus at Olympia. Perfectly proportioned, perfectly formed, noble and serene and, in a word, beautiful.

This idealisation of beauty was inherited by Rome and revived in the Renaissance. It is appropriate for the world of today, in an age where we are searching for formal arrangements and codes to create pleasing architecture and design. The early Greek philosophers tried to define aesthetic beauty through the geometrical laws of

proportion. They believed that if beauty is harmony and harmony is due to the observation of proportions, it seems reasonable to assume that these proportions are fixed.

This geometrical proportion was known as the Golden Section and has for centuries been regarded as the key to unlock the mysteries of aesthetic beauty. It was neatly summarised by Euclid in two of his propositions: "to cut a given straight line so that the rectangle contained by the whole and one of the segments is equal to the square in the remaining segment" and "to cut a given finite line in extreme and mean ratio." Simply translated the proposition states that a finite line should be divided so that the shorter part is to the

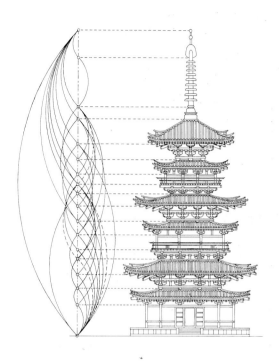

concept of unity and oneness we re arrange the ratio of a:b as b:(a+b) which develops into 1/Ø:1 as 1:Ø where Ø is known as the "Golden Proportion." It can also be identified in nature and be developed from basic forms and shapes of spatial order, to music and even to a sense of time. Through this development, the Abbeys of St. Bernard and Le Corbusier's La Tourette Monastry achieved not only simplicity and visual beauty but also through the reverberation of sound within the church, they transformed human chanting and singing into supposedly celestial music.

The use of the golden section is not exclusive to Western culture for there is evidence of its use in both the art and architecture of the East. In Japan, for example, the Pagoda of the Yakushiji Temple, which is known for its soaring grace and ingenious structural strength, employs the golden section to define the heights and dimensions of its six roofs. In a completely different way it can also be be discovered in the

Ryoanji Zen Temple garden. Although apparently at random, the rocks within the sea of gravel are located in relation to one another and to the garden environment in subtle proportions. The result is complete harmony between man and nature.

A common aesthetic appeal does exist. Dissatisfaction with instinctively personalised design, which is inhuman and incoherent, is also universal. The stigma of building something which is ugly and which provokes negative reaction can seriously damage the reputation of successful designers and the commercial success of a project.

The golden section is a good starting point in design development and a useful yardstick to assess visual coherence in design.

However, it is not the blind application of the golden section which will result in coherent designs and satisfactory forms, but rather the intelligent and calculated use of its properties with good judgement, that produces works of

longer part, what the longer part is to the whole. The resulting sections are roughly divided into ratios of 5:8, 8:13, 13:21 and so on. The ratios are never exact multiples. They are known in mathematical terms as irrational ratios.

Use is often made of the golden section today in proportioning the dimensions of windows and doors, picture frames, the pages of books and even the shape of a violin. The Pyramids of Egypt have been explained by it and Gothic Cathedrals interpreted by it by the relationship of transept length to the nave, of column to arch, of spire to tower and so on. So why not try it in the design of a bridge!

The Golden Section

In mathematical terms the golden section is the relationship between two ratios a:b compared with c:d. A simpler and more subtle relationship is a:b as b:c, and by introducing the

Far left. Golden sectioned Cistercian Abbey, Fontenay.

Left. Yakushiji Temple Pagoda, Japan.

Right. Golden proportioned spans and tower of the cable stayed A595 Duddon Bridge, Cumbria.

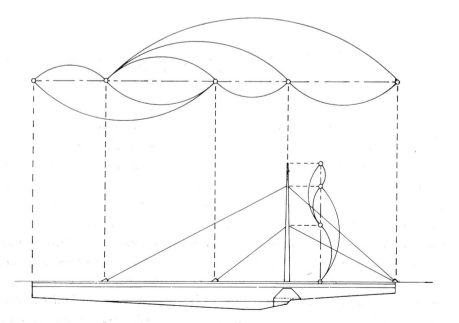

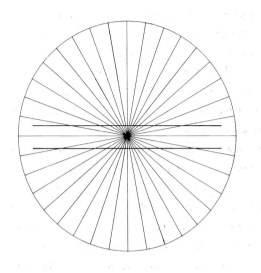

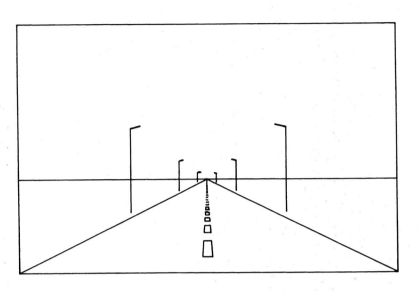

Far left. *The converging lines seem to make parallel lines bulge outwards.*

Left. *The parallel sides of the road apparently intersect on the horizon.*

art, architecture and engineering which will appeal to the senses and intellect. Aesthetic appreciation is dependent not only on dimensions and visual proportions, but also on light and shade, colour and texture, shape, form and setting. All these factors contribute to an impression on both a global and a human level.

Apparent and Objective Perception

We see the parallel sides of roads apparently intersecting each other at some distant point, but being parallel we know they do not intersect. Similarly converging lines to the parallel lines seem to make parallel lines bend outwards. When we see a line divided into two equal segments, a segment may appear unequal in size if the eye is made to travel quickly over the segment which is not disrupted by

the multiple partitions. Observation time along visual pathways becomes a factor in our perception.

These and many other phenomena are commonly called optical illusions and demonstrate the profound difference between geometric reality and visual reality. Classical architecture of both the East and West well understood visual perception and developed explicit rules to either enhance or counter these optical effects. Our senses detect and react to differences of proportion, not length or quantity. We perceive and measure the world around us through patterns of light and sound in real time and we invariably relate everything to and from ourselves.

This explains rhythm in architectural terms, which would be impossible if perception was not successive or time related. The ratios and proportions we perceive are durations of time and are not spatial quantities. It is therefore the proportions or the ratios between lines and

elements which makes them either expressive and dynamic or inexpressive and static.

Geometry and mathematics are rigorously precise whereas aesthetic ratios need only be approximate. The eye is not such a precise mathematical instrument. That is why precise ratios and millimetre accuracy have little meaning in terms of visual perception. The aesthetic ratios and proportions are the qualitative boundaries.

Theory into Practice

By looking at how these issues have been addressed in some of our recent bridge projects we can demonstrate how proportional consideration can be effective.

The overall composition of most major bridges will be dictated by technical requirements. For the A595 Duddon Bridge, a new bridge was to be built downstream of the existing one. The site characteristics of a bend in the river

and mature tree lined banks suggested a cable stayed arrangement with the tower located on a gravel bed on the eastern bank. The span and height of the tower were dimensioned to golden proportions.

In the design of long span cable supported bridges, the towers are the most prominent features. These elements require most careful consideration to ensure that they are pleasing in proportion by themselves and to the whole. Pylons need to have a clear aesthetic identity and yet be in harmony with the principal elements of the bridge.

The arrangement conceived for the cable stayed solution for the Ting Kau Bridge in Hong Kong involved a twin span triple towered pylon. The central tower, while it mirrored the outer two, had to be structurally much stiffer and therefore became the focus for the composition. In order to make the whole composition visually attractive the golden proportion was fully integrated into the design.

On the Tsing Ma suspension bridge, Hong Kong, detailed consideration was given to the tower arrangement

Left and below. *Golden proportions are fully integrated into the design of the triple towered cable stayed Ting Kau Bridge, Hong Kong.*

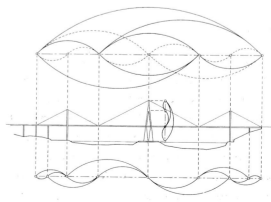

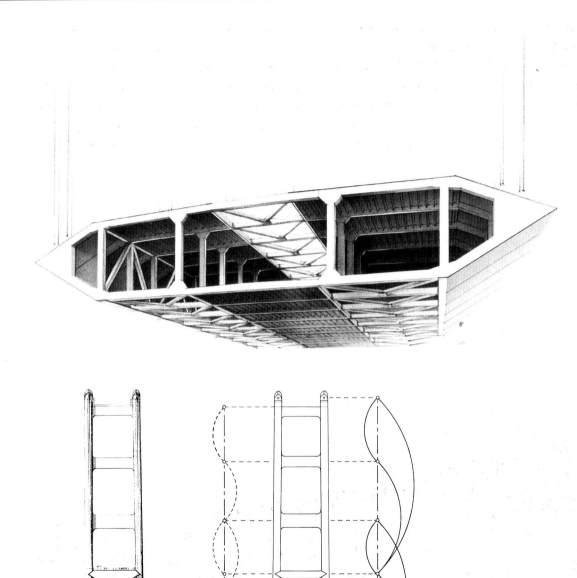

and to finishes around the towers. This ensured that the massive structures would not only be pleasing when viewed at close range but the overall harmony of the long span would also be enhanced. Judicious application of curves and a flare on the tower legs below deck level together with careful attention to the location of the cross beams between the legs were essential to soften the starkness of the towers. Considerable efforts were made to produce clear, light colours in the towers and other elements. In particular the saddles, which are massive structural items necessary to transfer forces to and through the towers, need to reflect their function and be in proportion to both the towers and the human form which observes them.

The double deck airfoil section of the Tsing Ma Bridge houses a high speed railway and two emergency carriageways within.

The pylons of the Tagus Road Crossing were conceived as pure sculpted forms, as if carved from a single piece. All

Above left. The golden proportioned double deck airfoil section of the Tsing Ma Bridge houses a high speed railway and two emergency carriageways within.

Far left. Elevations of the Tsing Ma towers.

Left. Proportions of the Tsing Ma towers.

Right. The Tsing Ma Bridge under construction.

surfaces flow one into another and clean simple lines emphasise the dramatic expression of their height. The pylon legs are cranked to allow vertical cable profiles and a sense of the concentration of forces being applied is expressed by a flare in the leg cross section below their change in angle. The portion of tower above is also subtly tapered to counteract the optical illusion created by the converging lines of the cable fan which would otherwise cause the structure to bulge visually.

Omission of a cross beam immediately below deck level is made possible by shaping of the base, clearly articulating the fact that the entire deck is actually suspended. The high level cross beam, necessary for seismic reasons, is carefully positioned according to golden proportions to complete the composition.

As we have seen, the use of proportion can be useful in the refinement of a structure. For the quality of our built environment to improve, our structures must begin to take on a human scale. However the blind application of the golden section, or any other system, cannot replace the creativity of the designer who sees and takes into account many other factors which cannot be easily quantified and defined.

The Expressive Fantasy of the Cable Stay

Leonardo Fernández Troyano and Javier Manterola Armisen, Engineers

It is difficult, at least for us, to analyse in retrospect what it was that led us to design a particular bridge, since every creative process is essentially intuitive and hence hard to rationalise.

Nevertheless there is one thing in the present analysis that does seem clear enough. Our way of understanding and shaping a structure has evolved over time, and it is within this ever changing framework where creative processes are placed.

Opposite. *The bridge Sancho el Mayor over the Ebro river in Navarra (Spain). It has a single central tower, three planes of stays, a box girder with transverse diaphragms.*

Far left. *The central tower and the back-stays, the cables open from the tower to the counterweights where they are anchored. The traffic flows below them.*

Left. *Upper part of the tower and planes of locked coil cable stays which are illuminated at night.*

The intention to give the leading expressive role to the stays in bridges comes from the almost magical regard cables have always had for us in structural engineering. The cable is the basic element in constructing efficiently designed long-span bridges. It is the most perfect and attractive component of structural engineering. The stay is the best example of any tension structure, where the visual expression of strength is immediate and obvious.

We will try to analyse the genesis of two cable stayed bridges, designed and constructed at different times. The cables in both bridges have been arranged in three fans, forming an angle of 120 degrees with the bridge deck. The spatial arrangement of the stay cables was designed to act with a single pylon located at one end of the bridge, holding up the entire span.

Sancho el Mayor Bridge in Castejón

The major north–south highway of Spain crosses the Ebro, the largest river in Spain, at Castejón in the Navarra region of Northern Spain. The width of the river at the point of the crossing is 100m. It can be bridged with a single span and eliminates the need for an excessively large bridge structure. The bed of the river is asymmetrical, the right bank is very steeply sloped while the opposite bank is quite flat with a large flood plain.

The asymmetry of the river bank suggested an asymmetrical bridge design. One main span will bridge the river bed stayed by a single pylon located on the steeply sloping side. The height of the road above the flood level of

the river on the shallow bank did not allow us to build a compensated span on this side, which is why we decided to use cable back stays for each side of the bridge deck, to also act like counterweights.

Starting from this initial idea, our next step was to define the geometry of the scheme, the deck, the pylon, the pier supports, the three fans of the cables and the counterweight blocks. We fixed the geometry as a direct consequence of the polygon of forces which act upon the bridge. We chose a tapered shape for the pylon because the inclination of the pylon presented the least force to support during construction and the profile looks well on the bridge.

The next step was to refine the concept design to respond to the functional requirement of the traffic lanes, the surroundings on which it was to be built and the intrinsic

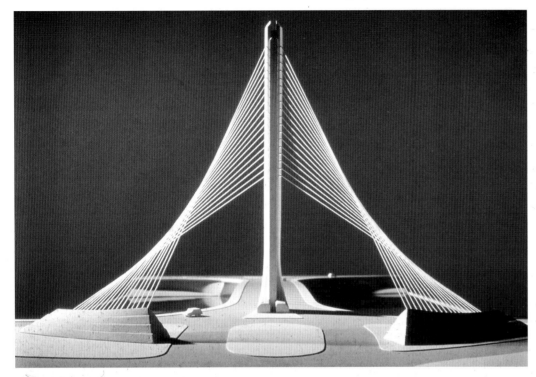

Above. Tower, back-stays and counterweights from the bridge over the Lerez river in Pontevedra (Spain). Due to their relative positions, the stays formed an hyperbolic paraboloid. The traffic interchange in this riverside is combined with the stays.

the reason the head of the pylon gets wider at the top when viewed along its longitudinal axis. In the other plane the pylon tapers as it rises, forming a 120 degree angle with the base. We studied two profiles for the pylon. One idea was to make a gradual transition which widens at the top and the other was to make an abrupt transition, with the top half of the pylon twice as wide as the bottom section, and each half having a constant cross section.

We decided that the gradual transition of the pylon head looked better and formalised the definitive shape of the bridge. Now we considered the cable stay arrangement which would be influenced by the type of stays and number to be used. When we were designing the bridge in 1977, a new type of stay cable was coming into use made from prestressing strand. However we decided to use locked coil strand cable because it was the best known solution at the time and also because we preferred it.

The lock coil strands could be bundled together to form a few mutually separate cable lines or used as single strands with a lot of cable lines. We chose the single strand stay as this is would make the cable fans quite dense. As a result the spatial disposition, the number and quality of the cable stays are the three contributing factors to the aesthetic design of the bridge.

The Lerez Bridge in Pontevedra

The name of Pontevedra is a transformation of Ponte Veteris, which means Old Bridge. In Roman times the town was called Duos Pontes meaning Two Bridges. History reminds us that this town has an affinity with bridges. We were aware from the very beginning that the design of the new bridge over the Lerez River would require a very special solution reflecting the link with the town, its colourful history and its bridge building tradition. The bridge design

strength of the bridge. This implies that good design solutions are conditioned by review rather than predetermined. Finding a better or worse solution will depend on the creative capacity of the engineer. The shape and dimension of the different elements of the structure have to be changed a countless number of times to meet the physical and visual qualities of the bridge and the crossing. Design *a priori*, in our view, never met the demands of the

site first time round. Perhaps the discipline imposed on the bridge by mother nature forces the engineer to design scrupulously to respect the physical world and the forces generated. It is one of the most stimulating yet exasperating features of bridge engineering.

The pylon of the bridge is conditioned by the width of the bridge deck at the base and by the need to hold the anchors of the cable stays at the head of the pylon. This is competition organised by the Xunta de Galicia emphasised this point.

The morphology of the river bed presented the same problems as the Sancho el Mayor Bridge over the Ebro. We were again dealing with an estuary slightly wider than 100m. However the surrounding streets were situated level with the top of the river bank and level with the bridge deck. The Lerez Bridge was not just a highway crossing like the Sancho el Mayor but also an important urban link to the town and to the streets on each side of the river. All this increased the expressive possibilities of the bridge, especially as it was going to be seen from many different standpoints.

The pylon of the bridge is located on a traffic island that is independent of the bridge deck. It is positioned a short way from the entrance to the bridge, in the middle of a dual carriageway that leads to the bridge. The dimension of the pylon therefore does not depend on the width of the bridge deck nor the gap of the central reservation of the deck. The longitudinal dimension and shape of the pylon was similar to the earlier design because the forces acting on it were similar. The pylon profile seen in the longitudinal direction was kept at a constant width, although the shape of the pylon was sculptured to emphasise the slenderness of the construction.

The counterweight blocks were placed on each side of the traffic island as though heralding the entry to the crossing. The blocks were laid almost orthogonally to the bridge axis and in such a way that makes the back stay cables form a hyperbolic paraboloid. We sought the greatest spatial expression of the cable fans and achieved this with a three fan arrangement and a three dimensional look to the back stays.

The hyperbolic paraboloid shape of the back stays changes in appearance depending from where you view the bridge. This was to be the principal attraction of the bridge, to be enjoyed by vehicle travellers going past the bridge or over it, and by pedestrians walking over it. Close attention was paid to the finished quality of the cable stays and to the number of cable fans. Since we designed the Sancho el Mayor Bridge the locked coil strand is no longer in use. Nevertheless the main span and back stay cables were arranged in 17 pairs. Not as many as the 35 pairs for the other bridge, but sufficient to reveal the hyperbolical paraboloid shape of the back stays.

We wanted to house each strand in a stainless steel tube which would then be cement grouted. We found that this provides the best finish to the cable stays. But cost reductions and new material technology has persuaded us to use a white protective sleeve to each cable strand of the Lerez Bridge. The quality of the finish is not as good as the stainless steel tubing but we hope they will prove to be just as durable.

The final work transcends the intention of the moment in which it was designed. The bridge begins to have its own personality. The scale, the surroundings and the materials used, all form a presence in which the design intent can be recognised. On some occassions the result makes a pleasing impression on us and in other times not so pleasing. We were pleased with both of our bridge designs.

Below. *The bridge over the Lerez river in Pontevedra (Spain) with a single tower and single box girder triangulated both internally and externally.*

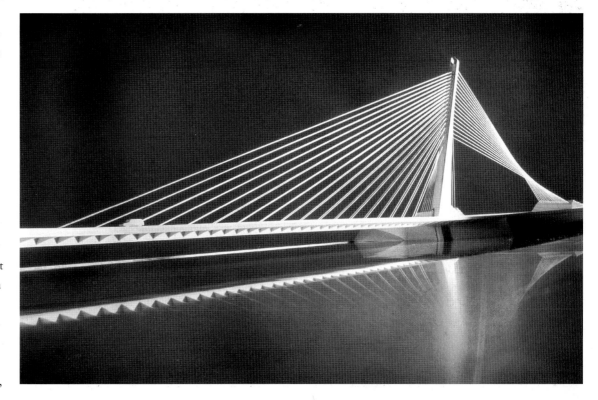

Twin Mast Bridge

James Carpenter Design Associates

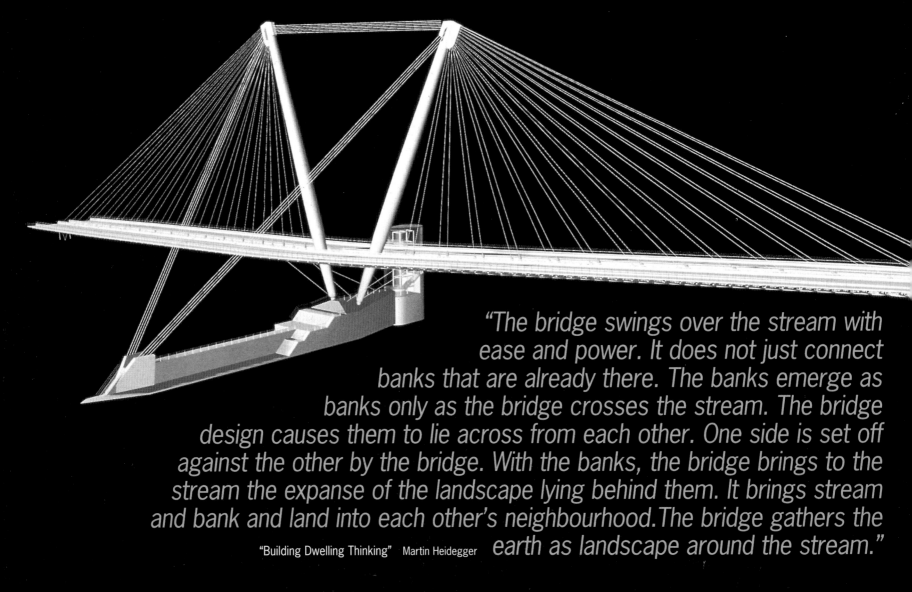

"The bridge swings over the stream with ease and power. It does not just connect banks that are already there. The banks emerge as banks only as the bridge crosses the stream. The bridge design causes them to lie across from each other. One side is set off against the other by the bridge. With the banks, the bridge brings to the stream the expanse of the landscape lying behind them. It brings stream and bank and land into each other's neighbourhood. The bridge gathers the earth as landscape around the stream."

"Building Dwelling Thinking" Martin Heidegger

The Crossing

The Mississippi emerges from a relatively narrow wooded gorge upstream to the wide and open valley at St Paul. The Native Americans called the place White Cliffs, and they considered Navy Island a sacred place. Many burial mounds still exist on the bluffs along the valley. Europeans made their first crossing of the upper Mississippi at the site of the present Wabasha Street Bridge and built the city of St Paul on the northern cliffs. The city boomed during the late 1800s as the primary transportation hub for the distribution of mid-west grain with barge and rail traffic passing beneath the northern span of the bridge. Light industrial buildings and abandoned industrial land dominate the southern bank. Nearby Harriet Island Park is the site of the annual construction of an Ice Palace, the focus of the Winter Festival celebrated in St Paul for over one hundred years. The Park, together with the mid-river Navy Island, is the urban component of a Great River Park envisaged as the focus for the city's efforts to revitalise the river valley.

The crossing is approximately 1200ft wide. The southern bluffs stand five hundred yards to the south, making

The bluffs.

Storming of the Ice Palace
Thursday, January 20th 1887.

the valley proportionately broad and shallow, a memory of the vast glacial river which formed this region eons ago.

The Brief

The existing bridge, built in 1889, was to be replaced with a new structure incorporating four 12ft wide traffic lanes, two 6ft bicycle lanes and two 12ft sidewalks for pedestrians and recreational cyclists. As the major vehicular and pedestrian crossing over the Mississippi River in downtown St Paul, its central location was significant as much as its historic importance as the first crossing of the river. Since the existing bridge had National Historic Structure designation, its replacement was regarded as a matter of great public importance.

The Constraints

Two railway tracks and a major road between the northern bluff and the river bank precluded construction in this area. The navigation channel passing under the northern half of the bridge had to accommodate a flotilla of barges 105ft wide and 1000ft long. A smaller road on the south side, with access to a marina for houseboats and recreational crafts, had to be preserved. Five hundred year floods, with the possibility of runaway barges at water levels 20ft above normal, were taken into account in the design of foundations and the clearance of the deck structure. Flight paths to the local airport had to be considered in finalising the overall height of the masts.

Initial Inspiration

The overall form of the bridge was inspired by the desire for a structure rising above the river, visible from the city and marking the end of Wabasha Street. This structure would signal the crossing from up and down river, establish clarity within the landscape, and recapture the overall scale of the river valley. The idea for the structure evolved from a tetrahedron created by the twin V-mast and the symmetrical cable stays fanning down either side of a straight roadway. The non alignment of the joining roads allowed the design of a curve in the bridge, with the cable fans tied to one side of the deck, marking the directionality of the river. The twin mast

Far left. Twin Arch Bridge, alternative design. Image: Macromedia Technologies Inc.

Left. Haunched Girder Bridge, alternative design. Image: Macromedia Technologies Inc.

Below. Site plan showing bridge alignment with Wabasha Street and the bend at Navy Island. Drawing: James Carpenter Design Associates

idea led to the development of a split roadway, which allows light to reach the river and parklands below the bridge, enriching the experience for both pedestrians and boaters at the water level.

The twin mast structure alludes to the erosion of the bluffs of the vast river valley. The cables supporting the bridge deck reinforce the city grid through the continuation of Wabasha Street, and provides a visual transition and hierarchy within the experience of the crossing. The design concept places primary emphasis on the mid-river island, establishing a superstructure rising from the island to support the road deck, clearly separating the pedestrian pathways from vehicular traffic.

In order to create a pedestrian experience of the landscape, the pedestrian path was originally designed to begin at street level in downtown St Paul and extend out at the level of the bluffs. The path would lead to the observation platforms at the roadway bend, high above the island, with elevator and stair access to get down to the island. From this gathering place sweeping views up and down the river are afforded to the pedestrian and the cyclist.

Qualities of the Bridge

The design character of the bridge was centred on the pedestrian pathways and aesthetic experience of walking the

bridge. This established the scope of work and our approach on design decisions. It also serves to extend the threshold of the city, to create an interstitial zone or boundary between the interior and exterior of the bridge. An earlier design for a glass footbridge revolved around heightening the experience of the water, landscape and light. It influenced our approach to the design of this much larger bridge.

The new bridge alignment is not perpendicular to the river and does not cross at the shortest point, but in so doing contributes to the overall integration of the bridge with its setting. The design also took the island topology into account, and maintained a clear unobstructed channel on both sides of the island.

From a great distance, the Twin Mast Bridge focuses one's eye on Navy Island, which is one of the unique features of the location and the only island within this dredged and man-made stretch of river. The twin masts provide a visual termination to Wabasha Street, and their asymmetrical stay cables, together with the bend in the bridge, acknowledge the flow of the river. The bend marks the location of the island to the passing motorist and the pedestrian, and is reinforced by the stair tower and glass elevator to the island. By dividing the north and south bound traffic lanes, daylight penetrates through the bridge deck to the island and the delicacy of the structure is emphasised. The stair and escalator tower bring the pedestrian

to the point where the masts meet the island. At night, the lighting of the stair tower creates a beacon in the landscape which illuminates the darkened river valley.

The mast foundation is designed as a broad staircase to the river's edge which provides a dock for pleasure boats. By concentrating the structure of the bridge on Navy Island, the need for massive foundations in the bluffs and on the flood plain is avoided. Great care was taken to concentrate the cables at the top of the masts in a fan shape to avoid excessive bending stresses in the mast, and to minimise the diameter of the masts. The detail of the cable connections was extremely important in reducing the size of the mast top.

Engineering

We asked Jörg Schlaich to clarify the structural adequacy of the twin mast scheme, the approach to detail, and to reassure us that it could be built. He was, in fact, surprised at the buildability of our idea. His extensive experience with cable stayed bridges and tensile structures became an important element in developing the concept to a finished design.

The deck structure consists of an open steel torsion box 20ft wide and 12ft deep, and cantilevered steel truss cross-girders at thirteen foot centres. These act compositely with the two concrete roadway slabs on top. The latter are separated by the width of the central truss box to allow light to

Far left. *Model of early design for pedestrian bridge and walkways at base of Twin Masts at Navy Island.*
Photo: David Sundberg

Left. *Night view of observation platform and stair tower.*
Photo: David Sundberg

penetrate through the bridge deck.

The deck structure is supported by two rows of stay cables anchored along the outer edge of the truss box at 40ft centres. The cables are attached to the tops of the two masts, one for each of the two halves of the bridge, which are angled at approximately 15 degrees in plan to each other. The bases of the masts are on the island, and incline at 22 degrees from the vertical toward their respective spans. The 290ft height of the masts is 20ft under the FAA flight restriction, making them lower than the towers on bluffs in downtown.

Two sets of six cables connect the two masts' tops and balance the horizontal components of the deck supporting cables. The lateral forces due to the curved roadway are transferred from the mast tops by two sets of cables down to a common foundation on the island. The lateral forces in the deck are also restrained by cables from the central truss box to the island immediately below the deck. There is no direct structural connection between the masts and the central truss box. Elimination of backstay cables allowed for smaller foundations at bridge ends and pivot joint details at mast base.

Process

The Wabasha Street Bridge commission was initiated by an international competition organised by Public Art of St Paul. We were selected from an international field of over seventy applicants. The brief emphasised pedestrian safety, views from the bridge, and access to Navy Island. The bridge was to be a civic symbol, the connection between the past and the future of the city.

Focusing on ideas developed in discussions with the Wabasha Bridge Task Force, we established the following goals for this project: to reactivate the long ignored river front, to emphasise and encourage pedestrian traffic across the river and to Navy Island, and to develop the bridge as a link in the pre-existing conceptual master plan for a linear park along the river. Most importantly, the structural expression of the bridge would represent the forward vision of the city, while the degree of detail would provide tactile enjoyment for the pedestrian.

We studied at least ten variations of the cable stay design before developing the Twin Mast Bridge. The logical solution to a typical cable stay bridge design would be a 2/3 to 1/3 double span with a mast on the west side of the island. The problems with this approach were that the masts would exceed FAA height restrictions and the span would be too low on the west side. More importantly such a design would not address the full width of the river valley.

We wanted the bridge supports to be at mid span for visual balance, and also to signal the mid span edge of the island, together with the overall width of the valley. A second variation on the cable stayed bridge was developed with a mast on the riverbank and pier supports from the island to the west bank. This design, which does not address the full width of the river valley, also places too much visual emphasis on the bluff.

Working with the Task Force engineers from St Paul's Public Works Department, and Jörg Schlaich of Stuttgart, we developed three alternative bridge solutions in order to qualify for federal funding. Although a steel arch bridge and a steel girder bridge were developed, we preferred the Twin Mast Bridge, and so did the client.

Above. Photomontage of Twin Mast Bridge in downtown St Paul.

Image: Macromedia Technologies Inc.

The relationship between two bridges, almost side by side spanning the same stretch of water, is like the relationship between two musical instruments playing a duet. The music sounds better if the two instruments are in harmony with each other and keep to the same rhythm and tempo. The two instruments should produce sounds that contrast, in order create a melody that is more interesting and colourful.

Årstaviken A New Rail Bridge for Stockholm

Foster and Partners, Architect

Simple Ideas Are Often The Best

Sweden is the place to enjoy rail travel. It is a country with a highly integrated public transport system, a policy that attempts to eliminate traffic congestion in the cities.

The policy has worked well because the rail network operates well. At regular intervals major improvements have to be made to the transport grid to match the expected expansion of commercial and residential development in the succeeding years.

When studying recent expansion proposals for the rail network, the management team took into account the traffic forecasts over the next 20 years. The study showed that the rail network to the south of Stockholm was expected to

double in capacity and this would become a serious bottle neck in less than a decade, unless improvements were made.

An additional rail route has been proposed in one of the most sensitive regions in Sweden, through the straits of Årsta, a fresh water inlet about 2km south of Stockholm. It is a beautiful spot and a much sought after residential area in a picture postcard landscape. If that wasn't enough to cause a public outcry, then the suggestion of building a new bridge without prior consultation right next to the famous "Årstabron" Bridge would be enough to start a riot. The Årstabron is to Sweden what the Forth Rail Bridge is to

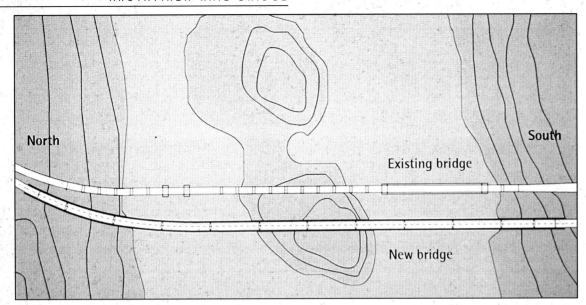

Scotland, the Sunshine Skyway is to Florida and the Golden Gate Bridge is to San Francisco.

The Swedish authorities decided that there was only one way to win public support for the new route in Årsta. Hold an international bridge design competition inviting some of the world's most respected designers and bridge engineers and see what happens.

This is the statement of our design for the Årsta Rail Bridge. We worked with Ove Arup International. It is based on a simple idea.

Duet for a Bridge

Some people question the involvement of architects in bridge design. Many respected bridge engineers all over the world have collaborated with architects to develop design ideas and to communicate ideas in symbols and words to convey the aesthetics of a design. They recognise that the training of an architect can bring a certain freshness to the design approach.

We collaborated with Ove Arup Engineers, as we have done on a number of other occasions. We worked as a team and not as a bunch of individuals with big egos trying to score design points over one another.

What follows is the text that we used in our competition document, with some additional notes to explain our design intent. The images we prepared were the real "vocabulary" of our submission.

We tried many versions. We decided that the best solutions were the simplest solutions. Anything complicated or dramatic would compete with the existing bridge.

We looked at many ways of solving the problem. Options included: replicate the form of the existing bridge, create three huge sickle arch spans, or create a sequence of flat arches with one large span for the shipping channel. All three solutions lacked a certain integrity and seemed only to respond to the existing bridge in either a deferential manner or an overpowering way.

The simplest form of a bridge for a railway is a concrete box beam bridge. There is no need for large spans since the largest shipping channel is only 40m wide. In addition the bridge could be built directly onto solid rock.

When we looked at the problem from first principles we asked ourselves what kind of bridge would we build today if the Årstabron did not exist and we were not trying to respond to it so self consciously? The answer was straightforward and uncomplicated—a box girder concrete bridge with regular spans about 75m to 90m apart. It would be economical to build and would meet all the site constraints, but would it respond well to the old bridge?

We chose the line of the bridge very carefully to avoid damage to the existing landscape. Only two supports are placed on the island of Alholmen which means that all the important trees can be preserved.

The proposed line of the new bridge also avoids the need to move Årsta Holmars Gard, an existing building on

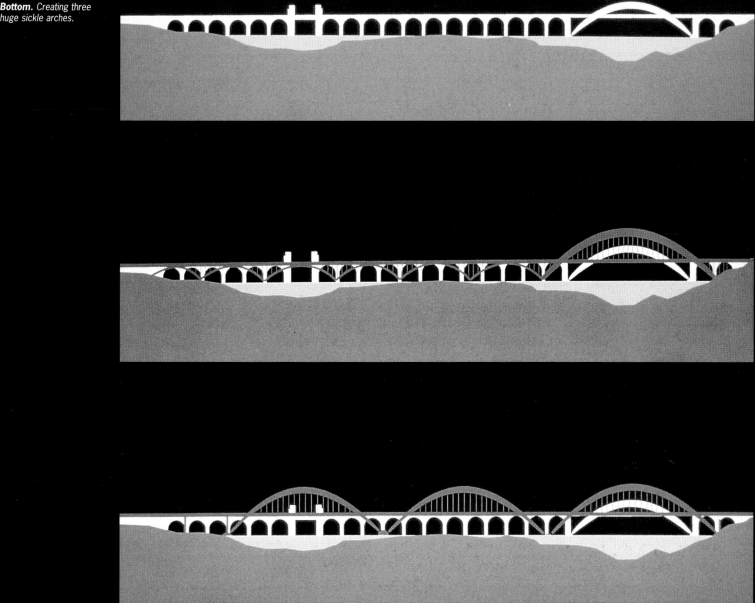

the island, as well as an old house known as Molitars Malmgard on the northern bank.

The line of the bridge is clear and simple keeping a respectful distance away from the existing bridge. At the same time it maintains the relationship between the two bridges.

When we put the basic profile of the new bridge against the Årstabron on a drawing, we were surprised to see how well they complemented each other. Certain refinements of the new bridge were still necessary. First the span lengths were altered to match the rhythms of the old bridge. Then the cross section of the standard box girder beam was altered to create a thinner profile when seen against the arches of the Årstabron. This was achieved by making the cross section a U shape, which deepened over the supports and tapered towards mid span.

The longitudinal section was modelled to create a continuous smooth profile, reminiscent of the hull of a giant catamaran. Coincidently the computer programme used to analyse the bridge deck structure was a programme used by naval architects to analyse ship hull designs.

Finally we researched the use of coloured concrete with the aim of setting up a contrast in colour between the two bridges. We chose a rusty red pigmented concrete, a colour which is typically used on many wooden buildings all over Sweden.

The bridge that we have designed can be seen in terms of a musical analogy.

The two bridges are in harmony with each other, they are parallel and at the same height.

They use the same rhythm of support spacing (although they are not of the same span). Yet they play different melodies and the instruments sound different (one bridge is grey, the other is red).

The two bridges work well together or apart.

The profile of the new bridge concept against the Årstabron.

Image of the new bridge with its hull shaped underbelly.

The design process has taken a relatively standard "bridge product"—a concrete box beam—and then refined it. The process of refining the new bridge was like tuning the strings of a musical instrument to find the harmony between it and another musical instrument, the old bridge.

This design process is very close to our mainstream architectural work where we regularly take standard components and assemblies and adapt them to the specific needs of the time and place of a project.

Our winning design for the Årstaviken is due to start construction in mid 1997.

Moving On
The Proposed Design of the Third Runnymede Crossing

Peter Ahrends, Ahrends Burton & Koralek, Architects

I felt that the bridge should be seen to be clearly and progressively modern and that, in the lineage of this family tree, there should be an identifiable sense of moving on.

In The Beginning

Symbolically the concept that is embodied by the word "bridge" is surely one of the most powerful and evocative in the vocabulary of our experience. The high flying movement of "sailing" eloquently but safely across a chasm, a waterway, or a crossflow that lies below is both profound and deeply satisfying. To make connections where none exist and to reach out and over to the other side lies at the very heart of our imagination and adventurous nature.

The Arup Bridge.

In designing any bridge we enter the territory of make believe, like a young child placing a stick across a puddle, we make fundamental gestures that speak and occasionally sing progressively about ourselves and our society. The design aim of the new Runnymede Bridge in my view was not simply to get the A30 diversion safely across the Thames at Egham. There was nothing particularly unusual about the span, the traffic volumes or any other typical engineering function of the bridge.

What was unusual was the site specific nature of the problem and the contextual significance of designing a bridge at Runnymede in relation to the two existing bridges. In this account of the design process I shall outline how these conditions and constraints were viewed and interpreted by the design team, and how these attitudes interacted during the design process.

Early Influences

In designing a bridge we reach forward to touch the future with our feet on the ground, rubbing shoulders with the past and our early influences. It is difficult, especially with hindsight and post hoc rationalisation, to cast one's mind back to the influential and complex patterns of thought that intermingled during the formative stages of design. However with this bridge certain issues were clear from the outset.

The Runnymede site is unusual not just because of the historic connections and natural beauty of a typical riverside scene close to the built up edge of Egham, but more importantly because of the presence of two fine bridges and the fact that they are to be joined by a newcomer. With the addition of the proposed new bridge, this group will be seen to be a family of bridges. A family that will share a number of common characteristics whilst having different types of construction and noticeably different forms of expression.

The first bridge is a steel arch bridge clad with warm red brick facings and stone dressings, designed by Lutyens and built in 1960. The second bridge formed part of the original M25 construction programme and was designed by Ove Arup & Partners as a set of arched white concrete frames and was completed in 1980. Sensibly this bridge was set apart from the Lutyens bridge by a gap of several metres. Not only did this assist with the new design and construction programme but it

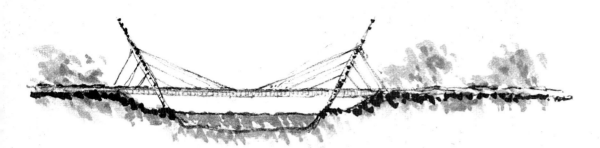

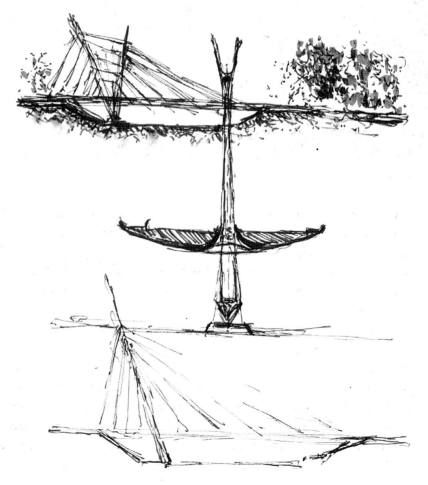

served to maintain and respect the design integrity of each bridge and allowed daylight to shine through the gaps to illuminate the river and towpath below.

Given these alternative idioms in which the Arup bridge has quietly established its own identity, the question became what should be the design options for the third bridge? In searching for solutions I wanted to identify those aspirations of the new bridge that would guide and inform our design intent in addition to satisfying the normal technical and budgetary requirements. Several ideas grew in my mind and these served as guidelines during the design development stage.

"The design should follow the earlier design precedent by establishing a gap between the new bridge and the existing Arup neighbour."

Indeed I thought why not go even further by separating the structure of the eastbound carriageway from the westbound lanes of the new deck, introducing a gap between them. This idea was also used in the separation of the footway structure from the traffic deck.

"It was important that the profiling of the underside of the bridge deck should be subtly curved in a monocoque form, in order to present the thinnest and lightest edge to the sky."

The structure should admit as much light as possible to the river below, which would be reflected upwards to light the bridge itself, so that it would be appreciated when seen from the towpath.

"The three bridges should be seen to progress from the earlier design expressed in brick and concrete and on to a modern interpretation in steel."

The Waiter's Fingers

Several options were considered. Some designs were presented to the client, some ideas were sketched for discussion between architect and engineer.

A cable stayed bridge was dismissed for three reasons. First there was doubt as to whether the mast and stays would be acceptable to the Civil Aviation Authority at Heathrow as radar interference was a possible problem. Second, from the engineering point of view, there was little justification for a cable stay in this position. Finally and more importantly, we felt that this solution was so obviously athletic and dynamic it would appear to be too self assertive and dominant in relation to the forms of the other bridges.

I thought the quiet relationship between father and mother might be disturbed by the decibel rating of a rather boisterous teenager.

Three options were considered, each being a structural separation on more or less a constant deck form.

The first was the simplest in formal and geometric terms. The two curves of the deck structure whose monocoque characteristics should have the inherent stiffness to span between two sets of inclined main struts in the form of an inverted tetrahedron. Frank Lloyd Wright's memorable description of "a waiter's fingers holding a tray" immediately comes to mind. Equally the deck should have the capacity to cantilever outwards from the primary structure to the leading edges to carry the pedestrian footpath. The scheme was developed, drawn, modelled in sketch form and presented to the Department of Transport, now the Highways Agency.

At first the client seemed less than convinced that this was the optimum solution. The underlying concepts were thoroughly argued and supported but some doubts remained. These concerned the inability of the bridge design to resist impact from boats and the ease of construction.

We considered a simple arched structure. But this design lacked some of the formal tension of the first alternative; a dynamic between the expressed primary elements and the monocoque deck. It was obvious that another move was called for.

The third and final option was typical of the building designs in our architectural practice. This concerns the celebration of the structure through the medium of daylight. Our design for the Cummins Engine factory in Shotts in 1980 illustrates this point.

Balance and Hierarchy

We proposed a more complex triangulated arched frame that serves as the spine of the structure and is situated in the daylit gap between the adjacent decks.

Cantilever struts extend out to establish a stiff edge to the monocoque, each side of the central spine. The dynamic tension of formal and structural interaction was once again re-established. The elemental characteristics of the parts became clear within an order of balance and hierarchy. Formal gestures and engineering checks seemed synonymous — one discipline seems to reinforce the other.

In keeping with the gap that we had designed between the

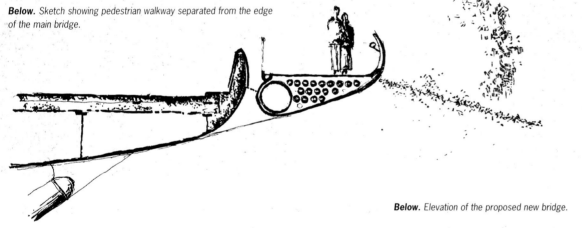

Below. Sketch showing pedestrian walkway separated from the edge of the main bridge.

Below. Elevation of the proposed new bridge.

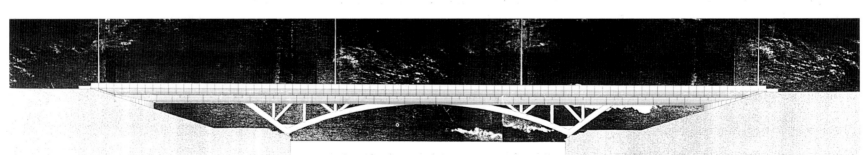

The first Runnymede crossing: the Lutyens Bridge.

deck structures, we introduced a second minor interval; a safe gap between the edge of the road deck and the footway. Cantilever beams established the separation and the connection. Happily this formal arrangement brought with it an unforseen bonus for it was at this stage that the services provision for mains gas and telecom cables became greater than had been anticipated. These elements were properly accommodated within the aerofoil section of the walkway structure. This enables easy access for maintenance and inspection, independent of the roadway.

Finally, a point of relative detail about the design of the abutment seen in relation to the form of the underbelly of the bridge deck. To contrast with the silver of the steelwork structure, we proposed that the abutments should be clad in stone, a contextual reference to the stone dressing on the Lutyens bridge. In relation to this the oblique and angular cut of the monocoque deck at each end of the bridge is intended as a formal gesture of recognition of the older bridge. It also addresses the conjunction between the elevated "feel" of the steel structure and the earth bound character of the stone-clad abutments.

It is here that the bridge will be seen to be rooted to the spot and at the same time free to fly across the river.

The outline design was submitted to the Royal Fine Art Commission and has been accepted. The scheme as a whole has still to go through the statutory procedures, and of course detailed engineering design, before it goes to tender. Changes could follow but I hope the design intent does not.

lifting, tilting & swing**bridges**

part *two*

Pont de Medoc A Swing Bridge for Bordeaux

Angus Low, Engineer

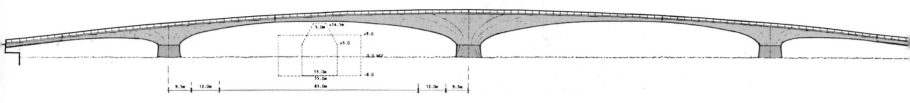

"When you are driving a car, you have to choose roads and follow roads and keep out of the way of other traffic. There is a lot of moment to moment activity dictated by the last moment and by the next moment. You are looking for signals and reacting to them. This is reactive thinking. So the walking-talking-breathing type of thinking is very like driving along a road. You read the signposts and make decisions. But you do not make the map. The other type of thinking has to do with map making. You explore the subject and make the map. You make the map in an objective and neutral fashion. To do this you must look broadly. This is quite different from just reacting to signposts as they appear." **Edward De Bono, from "Six Thinking Hats".**

Movement quite literally adds a whole new dimension to bridge design. The challenge is firstly to find the type of movement best suited to the circumstances and then to match the form of the structure to suit the movement. There are a surprising number of different movements to open bridges with:

swing - hinged lift - rolling lift - direct lift - retracting

Within each category there are many variations depending on the number of operating mechanisms that can be used. This diversity reflects the wide range of factors which affect choice, such as:

- the width of the opening
- the width of the roadway
- are there any approach spans?
- how quickly does the bridge have to open?
- how often does the bridge have to open?
- is it to be operated by the public or by specialist staff?

Good bridge solutions come from a thorough understanding of the specific situation and a careful sifting of all the possibilities.

Left. *Elevation of the bridge closed.*

Below. *Elevation of the bridge open.*

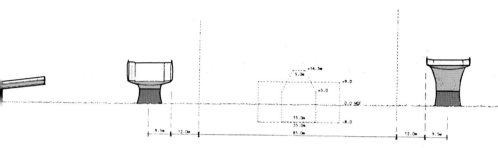

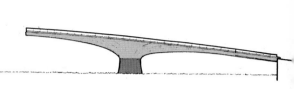

Broken Repose

At Bordeaux the swing bridge was required for large ocean going ships. A wide channel was needed to provide enough width to allow shipping to manoeuvre at low speed up and down the river. A channel 85m wide was specified with an additional 12m for a fendering zone each side of it, giving a clear span of 109m. The shipping channel was offset from the centre of the River Garonne, which was 400m wide at the bridging point.

This is big for an opening span, especially when a fast opening time of four minutes has to be taken into account. To achieve the required speed of opening, it is necessary to restrict the length of the swing, and this is why the opening span is split into two halves. Unfortunately, using two half spans results in an inefficient structure, which is not as structurally efficient as a whole span bridge in the closed position. To compensate for this the spans were made much stiffer, which meant a deeper cross section profile, particularly near the support piers. The supports were also strengthened to enable them to carry the greater cantilever

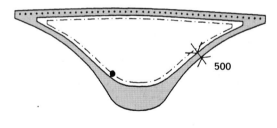

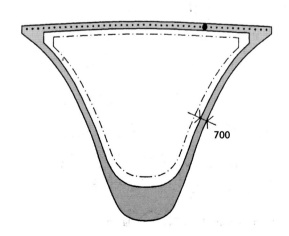

Above. *Cross sections of the deck indicating the sculptural form which is possible with concrete.*

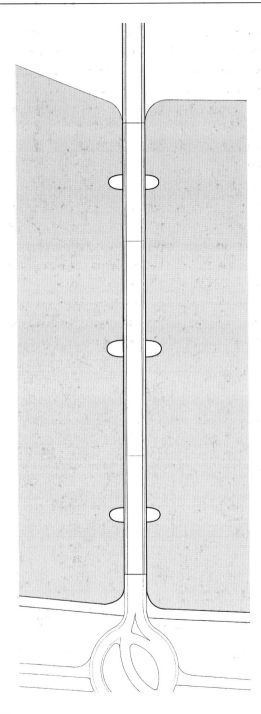

loads from the half span.

Swing action, with a horizontal motion about a vertical axis, is a natural choice for a large heavy span because it requires no vertical movement. It is particularly well suited for a multi-span bridge like the Pont de Medoc as the approach spans can be used as counter balancing arms which keep the opening sections stable when they swing. A vertical lifting span requires considerably more energy to raise, and when raised has to resist large wind forces. These factors influenced our decision for the half span swing bridge. It is worth noting that the length of the half span governs the structural depth of the span near the pier supports, which in turn is constrained by the maximum height of the roadway above the river. We wanted to make the bridge look slender and shaped the span to emphasise this quality, despite its rather squat structural depth.

The offset position of the navigation channel, and the resulting unequal split of the opening spans, gives this bridge a distinctive asymmetry. Under normal operating conditions the calm repose of the bridge is expressed by the symmetry of the spans. This disappears dramatically with the flick of a switch when the bridge spans are swivelled to open the shipping channel. The two end spans, which don't rotate, cantilever into fresh air. Equilibrium is restored just as quickly when the ship has passed through. The fact that one cantilever pair remains static while the other two swivel is all part of the surprise of the Pont de Medoc.

The effect could be described as "broken repose". The structural form which best reinforces this effect is one which emphasises this repose in its closed position. Hence we have looked for smooth, curved forms for the bridge profile to emphasise slenderness; using precast segmental concrete sections, sculpting the sections in two planes of

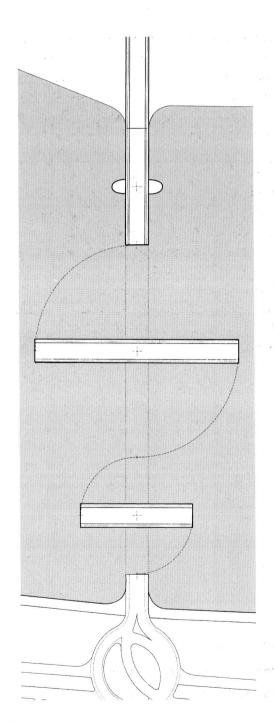

curvature. Steel would have been lighter but it just cannot be profiled in double curvature. The greater weight of concrete has been put to good use in the bridge deck, which acts as clamping force on the pier in the closed condition. This helps to absorb any out of balance forces acting on the cantilever spans.

Driving Force

Deep within, the swivel span piers are vertically mounted hydraulic rams which jack the bridge span slightly and act as pivots to rotate the bridge. However, just lifting and pivoting the span is not enough to cause rotation. A driving force is needed to make the span slew. The sheer size and weight of the spans, plus the fast operating speed, makes this requirement very onerous. Three options were reviewed: a horizontal ram drive, a vertical axis pinion wheel and a worm drive.

In general, hydraulic rams are simple to operate and control and are well suited to moving a large mass. However, no configuration of the ram drive could be found to allow a single ram to operate the span through 90 degrees. A more complicated two ram system would be necessary, which would require additional levels of control and co-ordination. A worm drive could be used, but it would be difficult to configure a large external drive wheel linked to the worm drive within the pier. We therefore chose the pinion wheel system for its simplicity and because its vertical axis drive seems appropriate for the slewing motion.

There was another important design issue which could impair the concept we put forward. It is to do with shear keys. Split spans operate in service with shear keys, interlocking them together and keeping them level under all load conditions. The shear keys have to be retracted with a hydraulic ram before the bridge can be opened. In the past

such shear keys have given a lot of problems. They are sometimes too small for the dynamic forces they have to resist from the pounding of heavy lorries and simply fail to operate or break up. In this design a careful study has been made to ensure that the lessons of the past have been fully reconciled. A suitably conservative design of the shear keys and their hydraulic operation has been adopted for the Pont de Medoc.

The proposed bridge structure with its asymmetrical swivel spans has been modelled on classical lines. Respect for local context was achieved by respecting the scale of the city centre, preserving the view from and towards the nearby buildings facing the River Garonne itself, and by restricting the overall height of the structure.

We wanted to provide Bordeaux with a unique and distinctive symbol which would make a contribution to the rich and diverse fabric of the city.

Left. *The bridge in plan, closed and open.*

Right. *The vertical axis pinion arrangement.*

Far right. *The pivot system within the pier.*

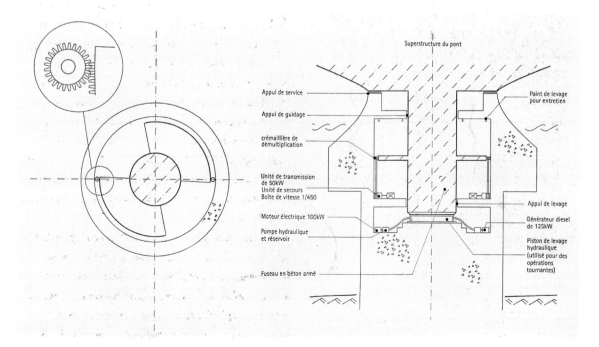

Bataille du Texel Dunkerque

Pascale Seurin, Architect

A metaphor of the link between town and port. A bridge isn't simply a functional structure but a response to a substantial urban concern. Highest or lowest, peripheral, central or middle, the main thing isn't the quality of the site, but where it is situated.

Above. The lifting mechanism consists of two 270t jacks. In storm conditions a strut locks the deck in an open position.

Left. Computer picture for the competition showing the masts in open and closed positions.

Below. On the sides, welded aluminium masks the box structure.

Opposite sequence. The bridge as urban action and symbol of revitalisation of a district in search of resurgence.

The construction of the "Bataille du Texel" lifting bridge in Dunkirk takes place in an important reconstruction programme designed by Richard Rogers, to expand the town centre towards the empty harbour which was left vacant by the closure of the shipyards. The bridge will link the Citadelle area on one side with the ancient Normed site, on an axis projected towards Malo-les-Bains.

The bridge sits at the head of the seaward entrance to Dunkirk and is the gateway to the urban project called Neptune which will develop the derelict port area.

"We didn't want to ask for technical plans for a bridge which could have joined onto the existing architectural project" explains Patrice Dordhain, in charge of the project on behalf of the Dunkerque Development Society. "We invited ideas for a bridge design, backed up by presentation sketches, to bring about a truly architectural statement for a bridge."

The total length of the crossing is 60m from bank to bank. The spans are divided into three, with a central lifting section of 28m and two fixed spans of 16m. This allows a central navigable passage of 20m. The lifting section is composed of two 14m long bascule structures. The cantilever decks are held together by a pair of monocoque tubular alumimium arms acting as robust edge supports to the composite steel and concrete deck structure.

What is unique about this lifting bridge is the distinctive

V-shape configurated by the elbow of the vertical mast and the monocoque edge beam of the deck structure. The lightweight fuselage mast structure is fabricated from extruded aluminium rib sections wrapped in aluminium sheets. The more robust edge beam arms have a similar aluminium profile but are much thicker in section and big enough to be inspected internally.

The symmetry of the V-formation and homogeneity of the structure accentuates the integration and loftiness of the bridge composition. The 15m high mast signifies the place of the crossing and the union between the port and the town. The use of prism shaped spot lights embedded into the mast piece to light up the structure adds the sculptural dimension to the structure by night, bringing together the urban setting of the site. It has a strategic position seen from Place du Minck; it is a place to meet and a metaphor of the close relationship between town and port.

Storm Position

The lifting structure supports a roadway 7m wide and a pavement 2m wide on either side. The raising mechanism is operated by 370mm diameter hydraulic piston jacks with a traction length of 2.9m. They are housed in the caisson foundation supports of the lifting bridge and are fixed to the heavily plated section on the underside of the deck structure. The stability of the lifting section is maintained by a set of tubular struts anchored to the caisson support foundation and attached to the deck structure by a pin connection.

When winds of 140km/hour blow up the Channel, the bridge is put into "storm position", with both decks lifted up. The structure is then locked by the thrust from a horizontal jack within the support caisson. The structure has been designed to resist wind speeds of 230km/hour.

"The stability of the bridge is generated by the forces on the

deck structure acting against each other to give static equilibrium when the deck is fully raised. Also the combined lateral resistance of the bridge supports, linked together by the framework of the bridge deck, is 240t" explains Maurice Souchay for constructors Baudin Chateauneuf.

The bridge can be opened or closed in 230 seconds or just under 4 minutes. The bridge control room looks like something out of Star Trek. As well as visual and audio signals the radar device in the control room can pick up the presence of a ship approaching the bridge from upstream or downstream, as well as the traffic build up on the roadway approaches. The signals are digitised, then relayed and displayed to the harbour control room monitor screens quite remote from the bridge site.

The client representative Patrice Dordhain refuses to fix an architectural on-cost for developing the sculptural and visual qualities of the bridge. He prefers to think of it as a relative cost to the urban quality of the Neptune project and something in the order of 5% of the cost of the bridge.

All photos: Phillipe Lesage
(Wormohout, France)

St Saviour's Dock Bridge

Bryn Bird, Engineer, Whitby and Bird; Nicholas Lacey, Architect

"A bridge establishes a new place."

Heidegger

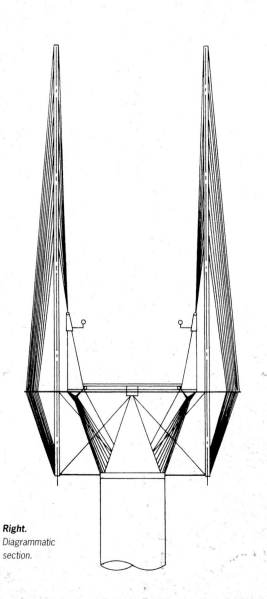

Right.
Diagrammatic section.

The bridge fills a gap in the riverside walk, and links together the two communities which were once divided by St Saviour's Dock. A filigree of steel, the design is a lightweight structure which relates visually to the motionless derrick cranes, and contrasts with the huge masonry masses of the historic warehouses fronting the dock: Butlers Wharf (and now the Design Museum) on one side and New Concordia and St Saviour's Wharves on the other; all reminders of the Dock's illustrious past.

This was a site ready for a fresh idea with its own active presence. We also wanted a design that would not become a barrier visually separating the Dock from the Thames or one building from another, and a bridge light enough to be swung by hand.

Lightness and transparency were achieved by turning to the principle of the bicycle wheel. In a bicycle wheel, the perimeter or the rim is a compression member that is kept

Below. *The site and view down St Saviour's Dock.*

in shape by the spokes of the wheel, the tension members. The idea is turned inside out. The main radial or compression members are the masts and tubular longitudinal bridge deck support. Arrays of circumferential stays are the tension members that bind the structure. Rigidity is enhanced by prestressing the stays. This is done by fully loading the bridge deck which slackens the lower array of cable stays. The lower array is then re-tightened and the load removed. The lower stays then react against the upper array thus tightening the whole. Within this tension system, there are other sophistications, such as the top horizontal compression member which is curved so that it has to be braced against buckling in only one direction.

The basic bridge structure weighs little more than a large saloon car, and is balanced about the pivot support using counterweights slung from the short end. A yacht wheel is utilised to turn the bridge structure about a horizontal plane, without much physical effort. An operating wheel is positioned on each side of the bridge.

To open the bridge, both entry gates to the bridge deck have to be shut. This releases the bridge and allows it to rotate freely when operated by the wheel from either side. Each wheel can be independently locked, so that if it is being operated from, say, the Butlers Wharf side it is a simple matter to walk over to the other side, lock the gate, then return to lock the gate on the operating side.

Stainless steel was used for the main structural members at low stress levels because of the need for welding which reduces limits on the material stresses. Vandal resistant cobalt steel rods were used for the main arrays.

The bridge as a whole comprises two timber jetties, two side-spans, new ramps on both sides and various extras to improve the security of local residents.

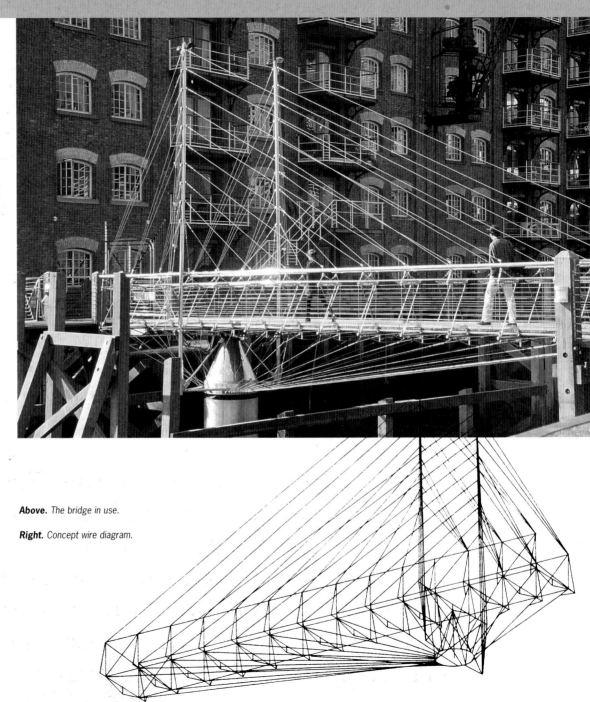

Above. The bridge in use.

Right. Concept wire diagram.

Dutch Rail's Double Track Lifting Bridges

R A M Steenhuis, Architect, H P van der Horst (Dordrecht Bridge), Construction Engineer, and L A van Hengstum, Technical Consultant (Gouwe Bridge), Holland Railconsult

If we wish to live, work and enjoy our leisure in a pleasant way in a densely populated country, then we must be careful how we allocate space. Dutch Rail wants to fulfil its future transportation task in the best possible way. We try to do everything to restrict to a minimum the disturbance to nature and the landscape surrounding future rail undertakings. Landscape and architecture are a primary concern to us. We investigate how a new rail track, viaduct or lifting bridge will best fit into the landscape and with least disturbance to flora and fauna of the region.

Below. *The Vuursalamander (translated firelizard) which only lives in a small forest in the south of Holland (the largest population in Western Europe) was threatened by realignment of railway tracks. It was saved by creating a special ditch next to the tracks.*

Middle. *A Dutch Rail train passing the flora and fauna next to the tracks.*

Bottom. *The railway bridge (right) with control room in the shape of a coffee filter and the road bridge across the Oude Maas, Dordrecht.*

The New Age of Rail

Spurred on by the overcapacity of the road network and noise pollution that knows no bounds, Dutch Rail is carrying out a thorough improvement to the rail network with Government backing, to become the primary transport system for conveying people and cargo by the turn of the century. The investment is huge, some 2.5 billion guilder per year or one million guilder every working hour for 15 years! But the payback to the people and the environment from the reduction in pollution and noise, the ease and comfort of travel and the frequency and capacity of the rail system, is worth the expenditure.

By the year 2000 Dutch Rail will be able to transport twice as many people by rail and carry three times as much tonnage in cargo, from 20 to 65 million tonnes per annum. Achieving this improvement to the rail network cost effectively has not been a case of laying lots of new rail routes nor upgrading all intercity lines with high speed tracks, but rather one of reorganising the management of the existing rail network and doubling the capacity by running double decker coaches.

The new rail system will offer a better inter city service whose efficiency will be improved by eliminating inter-regional train stops, bottlenecks, and restrictions over level and bridge crossings, rather than increasing the rail speed. The inter-regional network will be transformed so as to ensure short radius connections and a local service connecting the city centre with the outlying suburbs.

In smoothing out the bottlenecks to allow the rail network to carry more trains, many junctions have had to be upgraded. The two examples illustrated here fall within this improvement category. The first, the Gouda–Moodrecht junction, is characterised by the confluence of a number of lines, numerous road and motorway routes and the presence of a busy waterway, the River Gouwe, at the critical crossing point. The second, the Rotterdam–Dordtrecht rail crossing over the Oude Maas river, has been a major bottleneck for many years on one of the busiest routes in the Netherlands. Around 350 trains use this crossing every day, whilst the Oude Maas itself is one of the most heavily navigated waterways in Europe.

Hoist Bridge over the Gouwe

The existing steel swing bridge carrying traffic over the River Gouwe was woefully undersized due to its limited capacity of accommodating just two rail tracks. There were continuous disruptions caused to the railway network with the opening of the bridge for ship movement almost every hour. It was clear that a new hoist bridge with four tracks had to be built and that opening for passing ships had to be better regulated and limited to three times a day.

The problem was solved by Holland Railconsult constructing a new crossing point a short distance upstream, with new approach viaducts and a new vertical hoist bridge with four tracks. However, such a structure was very imposing on the flatness of the landscape and presented a high risk to the environment. It was important that the physical presence of the new bridge should not disturb or unbalance the harmonious image of the Gouda Plain and nearby town of Gouda. It was evident that the dramatic scale of the structure was such that the bridge would seek to "seize" the sky.

The 25m wide bridge is comprised of four precast concrete towers 45m tall with twin movable platforms 7m above the water line when closed. The platforms can be hoisted 34m high when fully open. Despite the imposing height of the towers, they have been profiled to appear slender and light weight from whatever point they are observed. Their vertical lines are accentuated by the curves of the tower face; featured portholes break up the bulkiness on the towers, while construction holes running up the face express height.

The four towers dominate the pair of lifting girders housing the hoist and the two cylindrical coupling beams that run parallel to the track. The movable bridge section is built up from a series of interconnected box beams that supports a composite steel deck section onto which the rail tracks are fixed.

Vertical lifting was found to be the quickest and most economical of all the lifting systems considered, from bascule to horizontal slip and to four track swing bridge options. It was the option that demanded the least and yet proved the easiest for maintenance. Two separate and independent movable platforms were chosen in order to make sure that one of them was available for railway traffic, should a ship accidentally shunt into the other or should there be a malfunction.

The massive concrete approach piers supporting the deck of the viaduct were hollowed out in the centre and rounded at the ends to create a pleasing aspect and to lead the eye away from the monotony of the deck line. Along the length of the viaduct a series of half moon shapes protrude from below the deck, amplifying the rhythm of the pier spacing and the point of the beam supports. The sensuous Brancusi like hollows and curves of the piers mirror the portholes, the curvature and the monumental mass of

the tower pieces. *"Looking through the circular hollows formed in the heart of the viaduct is probably like looking into infinity through a time tunnel framed in a curtain wall of concrete."*

All 36 precast concrete tower segments, each weighing 75t, were hauled up river on barges. They were hoisted off the barges by giant derrick cranes located on the deck of the approach viaduct, and then slowly dropped into position, one on top of the other. The design team had the choice of constructing the towers and the approach piers with in situ concrete. Using climbing forms the towers could have been

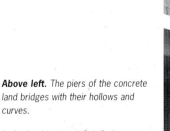

Above left. *The piers of the concrete land bridges with their hollows and curves.*

Left. *"Looking into infinity", the massive piers of the land bridges with the circular hollows.*

Right. *The double track lifting bridge across the Gouwe with two separate and independent moveable platforms.*

cast with an acceptable concrete finish and designed as a monolithic structure. However, the preciseness of precast production was a guarantee of achieving the near prefect plumb required for the tower and for operating the lifting mechanism. By precasting the towers the maximum out of plumb measured from top to toe was never more than 17mm, with each 75t segment built to a 2mm dimensional tolerance.

The Gouwe River Rail Lifting Bridge won the EC Best Concrete Bridge Category Award in 1992.

The Colossus over the Oude Maas

The towering white lifting portals of the Oude Maas Bridge near Dordrecht dominate the landscape like a modern colossus. For this span and for this location, the four track lifting bridge and the portal structure had to be fabricated entirely from high grade steel. The wider span of 55m over the River Oude Maas and the lifting height of 35m, meant a saving of nine million guilders over an alternative construction. The design of the lifting portals, the lattice bridge beams and viaduct spans was completed in less than a year. Holland Railconsult could have built a solid, grey, unintelligible edifice, reminiscent of so many heavily engineered structures that bestride rivers. But they did not.

The tapering trapezoidal profile of the portal legs, the smooth lines and the proportions of the coupling beams increase the grace and lightness of the structural form, with minimal need to dress the structure up behind a false facade. How many lifting bridges or barrages have witnessed such a concerted effort to model the aesthetic lines of the structure? Holland Railconsult involve and employ talented architects to develop their conceptual ideas in order to ensure that aesthetic sensitivity really adds to the quality of bridge architecture, rather than excusing such excesses as facile and a waste of engineering money. To Dutch Rail there is a price to be considered for imposing such a large amorphous mass on the horizon of an established landscape. Moreover, aesthetic consideration of design at concept stage will hardly show on the contract price. It is a price worth paying.

The Oude Maas lifting bridge consists of a pair of lifting portals and lifting beams and two double track movable bridges. The lifting portals stand on four ball hinges, one beneath each portal leg. The portal frame is made of a pair of 60m high trapezoidal steel sections—each weighing 300t—tied at the top by a 7.5m deep lifting beam. The portal frames are linked together by two 5.5m diameter coupling beams. A number of load cases were analysed using a space frame computer programme before the details of the lifting portals were finalised. Wind tunnel tests were carried out to provide a better understanding of the wind load factors. The lifting portal structure derives its rigidity from the joints between the various structural members.

Due to the fact that the movable bridge is balanced, the lifting portal has to carry the weight of the counterweights and the bridge deck. The movable bridges are balanced by the counterweights so that only a small excess load is needed to open or close them. At the base section of the portal leg a thick walled tube is welded to standing partitions within the

Below. The lifting bridge across the Oude Maas, completed.

trapezoidal section, creating a star shaped rigid construction which carries the force from the portals onto the ball hinge at the base.

The lifting bridges are driven independently of one another by two 132kW motors installed in the middle of the coupling beams. One motor runs four cable drums for the lifting of the bridge. On each drum there are two cables, one connected to the counterweight and the other one to the bridge deck. When the bridge is raised, the lifting cables of the bridge are wound up and that of the counterweight wound down. The lifting of the bridge is controlled by sensors. The bridge moves at 7m/second under normal power and 0.175m /second under emergency power.

Assembly to Order

The lifting beams and portal legs were split into three sections and the coupling beams into seventeen rings, each 3.1 m long, in order to facilitate factory fabrication. All the sections were fabricated in Belgium and then transported to Antwerp harbour where they were assembled into larger components and shipped by pontoon to Dordrecht along the Oude Maas. Four temporary gantry towers were erected on site to support the first two legs of the lifting portal and to carry the weight of the lifting beam. Once the first portal was positioned, the second portal and the tubes were fixed to the lifting beam before the temporary support gantry was removed.

The erection of the lifting portal and lifting beam was completed within the space of nine days. Some 550 bolts were required per joint, not only to transfer the forces between sections, but also to squeeze the joint plates together to prevent water from seeping in.

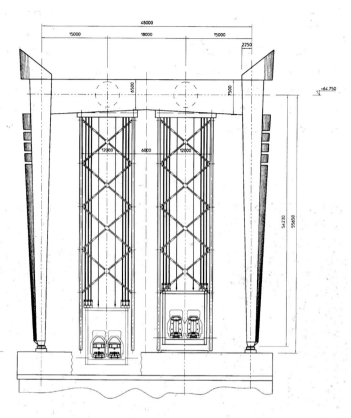

Above. The temporary construction together with parts of the lifting portal.   *Above*. The dimensions (in mm) of the lifting portal.

The principal components of each of the 88m span girder bridges — the lattice girder and floor section — were transported by road and by water to Antwerp harbour. From there they were floated on pontoon up the Oude Maas and brought to Dordrecht. The lattice bridge was assembled in two sections, one piece measured 104m long and weighed 930t, the other was 72m long and weighed 670t. The lattice sections were lowered onto the bridge piers by three derrick cranes. The main girders were then bolted and welded to the flooring structure. The new bridge was ready to receive rail traffic once the railtrack was ballasted and laid down on the deck structure.

Bridges, tunnels, viaducts and flyovers constitute important parts of a railway. Without construction work no train would run, it is just that simple. Deciding the optimum structure to incorporate into a railway is far less simple. Should it be a steel bridge or is concrete preferred? What foundation do we choose for the structure? A railway system consists of a track, the ballast bed, the sleepers and the rails. At first glance a relatively simple concept, but appearances can be deceptive. A deviation of just 2mm in the height, as thick as the lead in a pencil, causes a rough ride. Added to this is the fact the trains are running at ever higher speeds and greater axle loads. The design team chosen must be capable of giving first class design advice so that the structure harmonises with the environment.

A Cable Stayed Swing Bridge

James Eyre, Architect

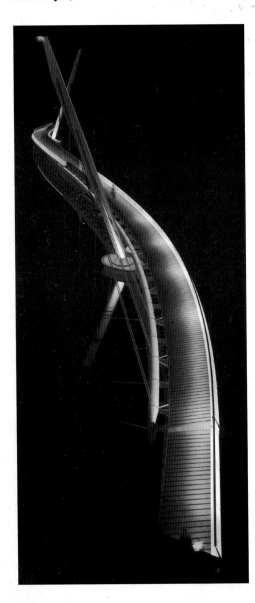

It is difficult to trace the exact inspiration for the "S" curve in architecture, in art and in science. A particular reference to the "S" curve was made in William Hogarth's "The Analysis of Beauty" where he compared the aesthetic qualities of various "S" shapes. He concluded that a particular "S" shape was the most beautiful, and referred to it as the "line of beauty". There are countless analogous examples in nature, ranging from the human form to the topography of sand dunes and the motion of waves. The fully evolved design finds this form.

Creative Instincts

The design for the South Dock Bridge crossing closely follows the very first idea that we mooted. It is unusual to arrive at the right solution so early in the brainstorming process, when architectural and engineering thoughts are explored without inhibition. It was a surprise because the design competition brief to build a footbridge across West India Dock to South Quay was very complex and quite demanding.

A footbridge was required across a 140m stretch of water; a part of it was to be temporary and a part of it permanent. But that was not all. The permanent part had to be designed so that it could be moved and relocated 90m east of its present position. Moreover, it had to have an opening section that would provide a passage for boats, 11m wide with a vertical clearance of 8m. When closed the bridge must allow a freeboard of 3m above the waterline.

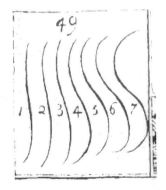

Right. William Hogarth's "The Analysis of Beauty" [1753]. The "line of beauty" is line 4.

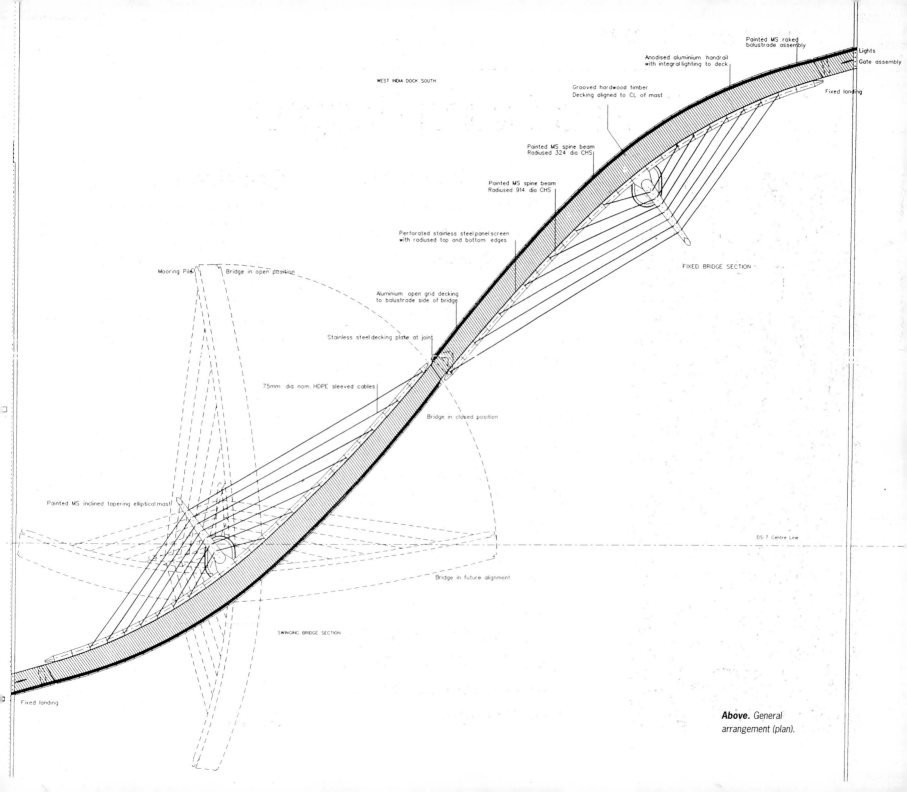

WEST INDIA DOCK SOUTH

Painted MS raked
balustrade assembly

Anodised aluminium handrail
with integral lighting to deck

Lights

Gate assembly

Grooved hardwood timber
Decking aligned to CL of mast

Fixed landing

Painted MS spine beam
Radiused 324 dia CHS

Painted MS spine beam
Radiused 914 dia CHS

FIXED BRIDGE SECTION

Perforated stainless steel panel screen
with radiused top and bottom edges

Mooring Pile

Bridge in open position

Aluminium open grid decking
to balustrade side of bridge

Stainless steel decking plate at joint

Bridge in closed position

75mm dia nom. HDPE sleeved cables

DS-7 Centre Line

Painted MS inclined tapering elliptical mast

Bridge in future alignment

SWINGING BRIDGE SECTION

Fixed landing

Above. General
arrangement (plan).

Finally the deck structure had to be sufficiently wide and smoothly inclined to allow wheelchair access.

We were hesitant about running with our first idea, which suggested a pair of cable stayed bridges stretched diagonally across the dock. One of the cable masts was positioned so that it only had to rotate itself into final alignment sometime in the future. However, we were still unsure whether a cable stayed bridge could be rotated with a span of 80m. We also had to fix the landing position on South Quay in its future location, rather than the present one, and we were concerned that the client would reject this suggestion.

Having considered a wide range of other strictly conforming solutions, including floating pontoon-based designs and cable stayed structures, we kept coming back to the first idea because it solved all the bridging problems. Was there some unseen reason, we agonised, why the landing could not be in its final position from day one? All the indications were that the desired line for the bridge pointed towards a diagonal crossing connecting Heron Quays Station on the north side, to the core of development of South Quay. Just as

compelling was the news that a cable stayed swing bridge with a span of 80m was within the realms of practicality. So we decided to take our chances and run with the idea.

This approach was vindicated as we later found out, after we had won the competition. We were informed by the client that we had solved another bridge problem which was not in the brief. They had need of another new bridge further to the east, over the same stretch of water. Our proposal allowed them to reuse the redundant half span of the bridge, when Heron Quays was fully developed. They discovered that with our solution they were getting two bridges for the price of one!

Left. Design development sketch of deck construction.

Concrete in tube to Counterbalance

Timber deck

'Purlins' on Ribs

The Structure

The elegance of the bridge is derived from the relatively few structural elements that bring it together—the lines of the curved deck, the visual counterpoint of the raking masts and the play of the cable fans.

The appeal of the curved deck, lies in the compound nature of the "S" curve it follows. Also we increased the width of the deck towards the mast position and narrowed it as it approaches the centre of the span. This enhances the apparent length of the open span and the visual lightness of the structure. There is also a camber arrangement to the deck which, with the switchback configuration and the tilt of the masts, creates a dynamic interaction between the two halves of the bridge.

The bridge deck itself is supported by a curving spine beam—the boom—which is picked up by the cable stays suspended from the mast. Cantilevered rib sections spring from the spine beam to cradle the timber walkway and steel support joists that span between the ribs. The inner walkway edge is screened by a stainless steel perforated sheet, which will give shelter to pedestrians from brisk winds, while appearing transparent. We nicknamed it the metal hedgerow. The open nature also serves to ensure aerodynamic stability preventing harmful oscillations. The outer balustrade on the other hand, comprises a series of angled posts and bands of horizontal flats, which serve to reinforce the horizontal line of the deck.

The length of the cable mast was tapered and shaped like a rugby ball in section, but derived from a single radius tube of 1.8m diameter. We all agreed that it was critical to avoid a regular conical shape to the mast, as it would look dull and clichéd. It offered no directional focus or momentum due to its regularity of form.

Developing the details of the mast shape for fabrication was to prove difficult and expensive as it had to be made from 30mm thick steel plate. To achieve a vertical taper in the mast, the idealised egg shape would result in a double curvature along its flanks. It was therefore essential to find a form that could be fabricated with only a single curvature.

The compromise resulted in maintaining a circular shape at the base of the mast, with the remaining section made up of two cylindrical wedges that taper towards the top. This achieved the blade like form that we sought.

The curvature of the walkway and spine beam and the relationship to the raking mast introduced the possibility of out of plane cable stays. The twisting of the stays alters the perspective, giving the bridge a three dimensional quality when seen from any angle. We chose the regular harp cable arrangement rather than a fan array because it was more economic and the more minimalist statement.

The anchor points connecting the cables to the spine beams and to the mast are both eccentric to the centreline of the structural components. This reduces the torsion effects in the spine beam generated by the cantilevered deck.

The foundations for the masts are supported on driven steel piles which are capped by precast shells filled with concrete. A slew bearing was positioned at the base of the mast of the opening span over which the entire opening bridge section is balanced. The bearing is driven by an hydraulic motor mounted on top of the pile cap and enclosed by the maintenance access deck.

Left. *Section at mast position.*

Happy Coincidence

We formed the design team by chance rather than choice. Such luck has worked to our mutual good fortune, despite having no previous bridge design experience ourselves nor any experience of working together. We turned a chance encounter into a very successful partnership.

The client had drawn up a shortlist of engineers and invited them to work with an architect on a list they were issued. As one of the architects on the list we contacted the listed engineers that we knew, only to discover that all of the better known practices had already teamed up an architect. The Chris Wilkinson–Jan Bobrowski team emerged from the left overs on the list. What pleased us both was the immediate synergy and enthusiasm that we shared for the bridge competition. For both architect and engineer, left on their own, without the other's help, this design would not have been possible.

The combination of sculpted form and functional constraints of the bridge confirms the belief that architecture is a fundamental part of bridge design. Out of engineering necessity and the straightjacket of constructability, the chrysalis of a good idea can be transformed into a lyrical and romantic expression in a bridge that is both exciting, dynamic and economic to build.

This was the very first bridge competition we had entered. It was the very first bridge we had ever designed. It was great to be told we had won the competition.

Right. *A computer generated photomontage to show the dramatic intervention against the night view of Canary Wharf.*

A Balance of Shifting Masses Concept for a Lifting Footbridge

John Lamont, Engineer and Alain Spielmann, Architect

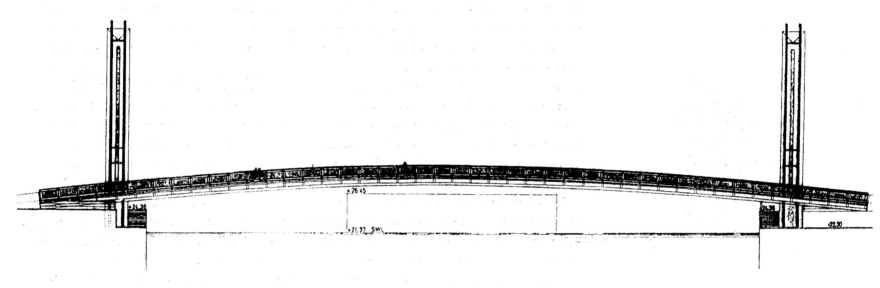

We have designed a simple yet elegant lifting footbridge whose aesthetic is a subtle dramatisation of a pedestrian waterway crossing, leading to and from the bright and dazzling architecture of a major New Arts and Leisure Complex. The bridge is a transition from calm to chaos, an escape to the quiet of open spaces and whispering trees on one side or an entry into the noise, the people, the smoke and the smell, of fun and laughter on the other.

Relationships and Relativity

The completed redevelopment of Salford Quays has been planned to coincide with the dawn of the new millennium. The remaining land area within this derelict dockside landscape has been earmarked for building a major new cultural centre for both the performing and visual arts. The centre piece of this £100 million development is the Lowry Centre, a £60 million project designed by the late Sir James Stirling and his successor Michael Wilford.

The Lowry Centre and its immediate surroundings will provide a landmark for the area and for Salford. The establishment of clear linkages between all the built elements, especially the new footbridge over the Manchester Ship Canal, is absolutely essential. The footbridge is in close proximity to the proposed Arts Centre and will be seen, from many vantage points, in front of the shifting mass of buildings.

A triangular plaza feature of the Arts Centre forms a significant new space within the complex and acts as a central focus to the surrounding areas and a generator for future

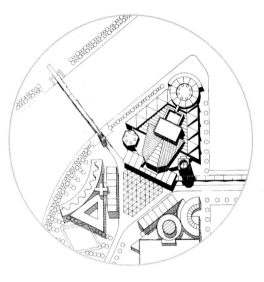

Below. *Plan of the Lowry Centre, Salford.*

development. The open corners of the plaza allows three axial approaches. One to the east describes the edge of the new parkland space on an axis with the entrance to the Lowry Arts Centre. The northern approach arrives adjacent to the new Metrolink station on the dockside bund and the prominent administration tower of the Centre. The southern approach is provided by the new footbridge which reaches across the ship canal to connect with Trafford Park and beyond.

The bridge is a vital pedestrian connection, as well as a significant composition within the new plaza and the Lowry Centre itself. The design of the bridge is also an expression of a functional bridge that has to be opened for ship navigation. Access along the canal bank, including access for boat moorings, landing points for a waterbus, bridle ways and towpath, as well as for the physically handicapped, has to be maintained.

Aesthetic Minimalism

In contrast with the massing and voluminous shapes of the Lowry Centre complex we have designed a simple yet elegant bridge whose counterpoint is to connect, unify and enhance the impact of the surrounding architecture. Lightness and transparency accented by visible clues to the integrity of the bridge construction is the basis of our design approach.

We see no merit in making a grandiose statement for this bridge in such a setting. Such a bridge would look out of place and out of balance with the massing of the new buildings seen by day and lit by night.

The mystique of the bridge in such a setting needs to be discovered rather than advertised. Its presence will be a sign post on the skyline, playing on curiosity from afar, but no more. As the bridge is approached, it begins to dominate the horizon and become the focal point with a character and charm all its own.

The bridge must be a minimalist expression of understatement, not a muscular metaphor. It seduces and informs of its subtlety, the more it is visited.

The flat parabola profile was found ideal for the bridge deck aesthetic, but the slim structural form of the deck was made to work hard to span 90m across the canal. The rising arc of the walkway elevation has been kept below the tops of the tree lined embankment either side of the canal, to enclose the space over the waterway. Seen from the bank, the tree lined waterside promenades block the visual intrusion from the Arts Centre buildings and Trafford Park, to give the bridge a canvas.

The lightness and quality of the structure is picked up by the unbroken line of the curving box girder support beam, the asymmetry of the cantilever walkway that springs from the girder beam and the sweep of the arc echoed in the handrail line.

Arriving at mid span, the pedestrian will experience the transition from tranquillity to excitement looking along the arch of the footbridge, at the lifting mast at the far end and onward towards the colour and dazzle of the Lowry Arts Centre itself. The stiletto shaped lifting tower and slim guide rail ladders positioned at the entry and exit points of the bridge signal the presence of the bridge and its location from afar.

A Fanfare for Simplicity

The bridge deck is supported by a steel box girder beam which is eccentric to the walkway centre line. It spans the canal in a shallow parabolic arc. The girder beam protrudes above the walkway on the upstream side to form the lower part of the parapet wall. The girder beam is supported at each end by lifting towers acting with the guide rail ladders. The guide rail and lifting towers are connected to each other at the top and the base, to form a braced structure that rests on a piled foundation cap.

The deck structure cantilevers from the main girder in a series of tapered beams forming ribs, to carry the steel joists which support the timber deck. On the down stream face the parapet and handrail is inclined inwards, towards the walkway. The deck is lit by low level lighting mounted on the underside of both parapet handrails.

When raised the main girder beam of the bridge hangs off two ogival shaped lifting towers located on the banks of the canal. No supports are located within the canal waterway. The circular guide rails are positioned to pass through the bridge deck, within the walkway section, to one side of the main girder beam. The cable mechanism and guide rail to raise the bridge is enclosed by cladding panels fitted around the lifting towers. The lift motor is housed at the base of the tower.

During raising and lowering of the bridge, the two ends of the bridge will be synchronised by precision encoders to ensure that the bridge remains level at all times. Multiple roller chains are employed to support each lifting carriage, the chains of which are suspended from chain wheels in the top of the lifting tower. In the fully lowered position the bridge will rest on separate bearings so that the lifting mechanism will be free of load when not in use. The lift motor drive is designed to optimise on the smooth acceleration and deceleration of bridge lifting. The bridge can be fully opened or closed within 3 minutes.

The lifting mechanism is based upon a proven system used on Royal Navy ships for aircraft lifts. They cantilever from their support tracks in a similar manner to that proposed for the Lowry Footbridge. Operation of the bridge is carried out from a kiosk that can be positioned remote from the bridge. The bridge control lever is spring loaded and therefore has to be held continuously to operate the lifting mechanism otherwise it will stop. There are separate switches for warning lights, waterway navigation lights, bridge deck lighting and for opening and closing the gates that lead onto the bridge deck.

Spot lights mounted at the tops of the towers create a merging pool of elliptical light over the length of the footbridge as well as along the axis line to the plaza.

Construction of the lifting tower foundations will be done from pontoons to avoid damage to the quay side or to the weak canal embankment. The tower steelwork will arrive on site fabricated in sections to be lifted and welded in place using barge mounted cranes. The box girder deck beam will be brought to site in sections then welded together on a barge, before being floated into position using barge mounted cranes. The girder beam will be erected on a gantry framework built on the barge. The girder will be raised clear of the towers using jacks and then lowered down over the guide rails suspended from jacks located on the top of the towers.

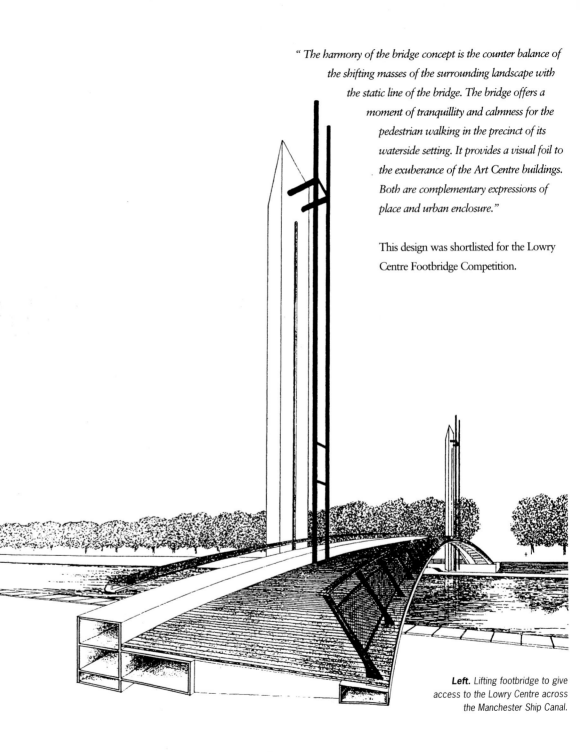

" *The harmony of the bridge concept is the counter balance of the shifting masses of the surrounding landscape with the static line of the bridge. The bridge offers a moment of tranquillity and calmness for the pedestrian walking in the precinct of its waterside setting. It provides a visual foil to the exuberance of the Art Centre buildings. Both are complementary expressions of place and urban enclosure.*"

This design was shortlisted for the Lowry Centre Footbridge Competition.

Left. *Lifting footbridge to give access to the Lowry Centre across the Manchester Ship Canal.*

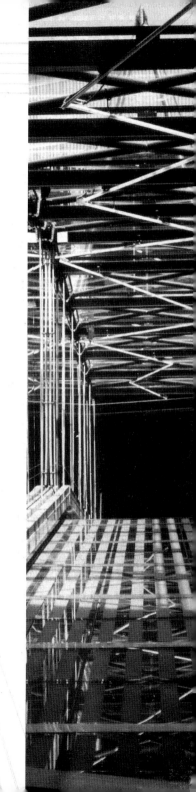

community**bridges**

part *three*

The Great Arch Over the Danube An Idea for World Expo '96

Rodolphe Luscher and Filippo Broggini, Architects

Buda the historic, Pest the merchant and Obuda the provincial were fused together in 1872, in the name of a great city. She swept away her old quarters and in their place built her bridges which have since become the jewels of her architecture and much admired the world over. The first underground railway in Europe was built here. The first lattice bridge could soon follow.

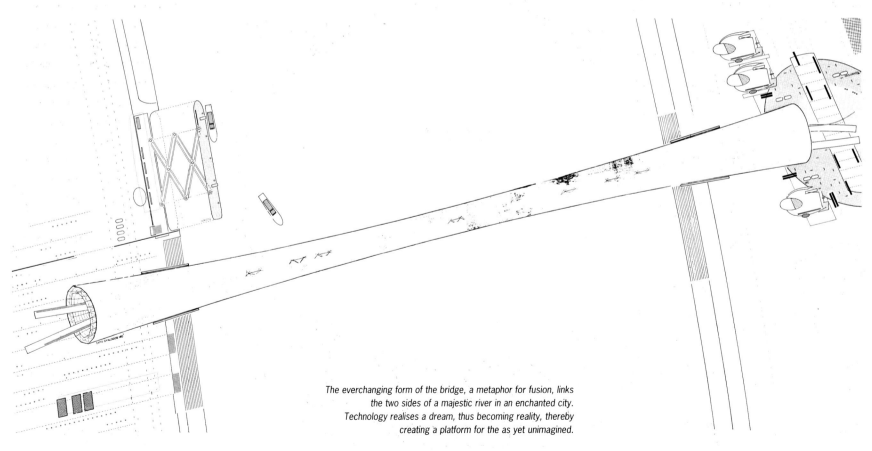

The everchanging form of the bridge, a metaphor for fusion, links the two sides of a majestic river in an enchanted city. Technology realises a dream, thus becoming reality, thereby creating a platform for the as yet unimagined.

Leaping The Danube

The bridge is a case of extending the tradition of great bridge building in Budapest along the banks of the Danube which started in the middle of the 19th century. Technological innovations, big spans and monumental structures have competed with one another for over a hundred years. After the arched bridge came the post and beam bridge, then the suspension bridge, the box girder bridge and the cable stay bridge. We now introduce the lattice bridge, a cat's cradle like structure, capable of leaping across the 500m of the legendary waters of the Danube.

Functionally the arch, in contrast with the surrounding rail and road traffic bridges of the 19th and 20th centuries, would be erected to the glory of the pedestrian and for "homo communicans" of the 21st century. Inaugurated during the 1996 World Expo, the arch could take on the role of a majestic "pontifical" entrance capable of organising,

Left. Three dimensional portion of the lattice structure.

Below. Schematic structural cross section illustrating the links and nodes principle of the lattice structure—a metaphoric transport map.

regulating, accelerating, slowing or reversing the flow and flux of visitors at any time of the day.

Following her duties at Expo, the bridge will become a vast welcoming structure with modular spaces for helping to commercialise the enterprise of Expo, long after it has gone. The economical cost of building the arch could be offset by the potential revenue generated from the cafés, restaurants, galleries, shops and other gossip venues. The structure is a metaphor for the trellis, the vine pergola and the god Bacchus symbolising conviviality, excess and enjoyment. Ecologically the arch reveals itself as a departure point for the regeneration of urban routes and pedestrian networks within a city. Two principles adopted for the project must be retained to stimulate the germ for ecological conquest of the city. Man's fight against contradictory values of urban disorder and the car economy. The arch stands for the link between the green spaces and the conveyer belt society.

The bridge links the park lands on one side to the newly created square at the Pest at the entry point to the bridge. It is also a conveyer footpath, a viable substitute for the car and a compliment to the city transit system.

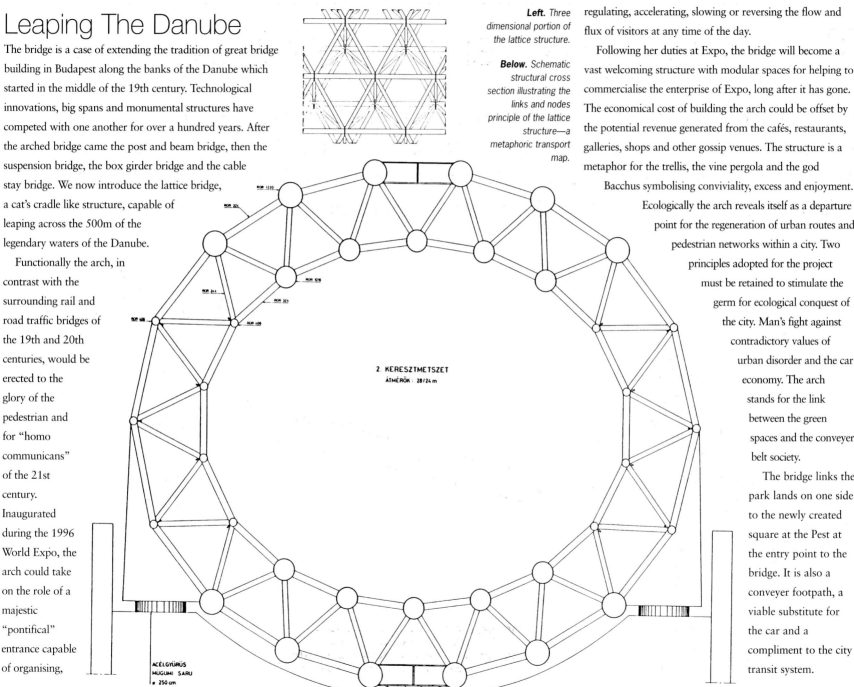

2. KERESZTMETSZET
ÁTMÉRŐK : 28/24 m

ACÉLGYŰRŰS MŰGUMI SARU
ø 250 cm

VEZETŐ SARU

A Symbol For Humanity

I would like to think that the arch expresses as much the theme of Expo '96 as it does the rebirth of the Hungarian nation. At the last millennium the city hosted the construction of a great monument at the city gates, the Hosak Tere. For the next one it is the Expo Bridge implanted in the heart of the city, bridging the cultures of eastern and western Europe.

The bridge is like a Noah's Ark navigating the stormy waters of contemporary ideologies, economies and conflicts. It symbolises by its social context and strategic location all the necessary elements for the cyclical restoration of yesterday's, today's and tomorrow's urban life in Budapest.

But even more than the symbolism of the bridge and the testing ground between opposite banks, here in the arch is the university and technological park of scientific research and discovery forging a universal and eternal alliance between past and present, and promoting the Budapest tradition of café life, the baths and good humour.

"Is it that the river is reflected in the symbol of the arch or is it the reality of the bridge which loses itself in the flowing waters of the river?"

Winner of the Budapest Expo '96 Bridge Competition.

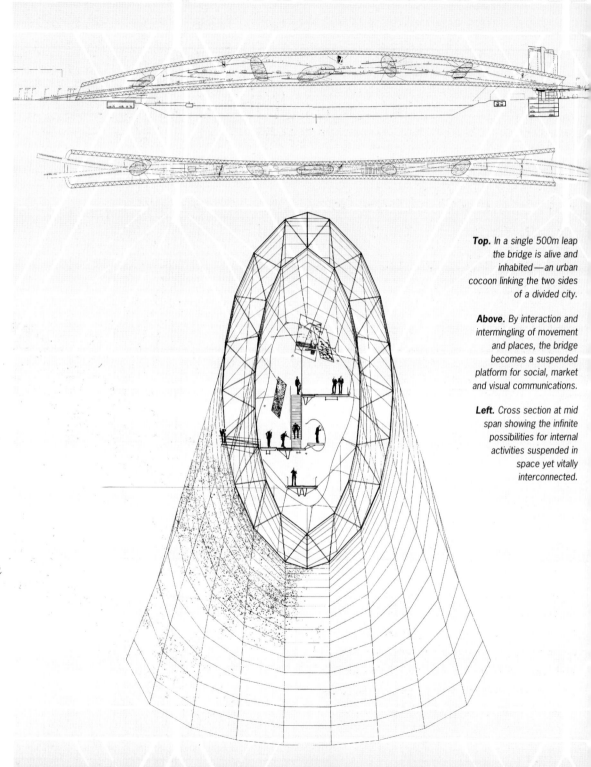

Top. In a single 500m leap the bridge is alive and inhabited—an urban cocoon linking the two sides of a divided city.

Above. By interaction and intermingling of movement and places, the bridge becomes a suspended platform for social, market and visual communications.

Left. Cross section at mid span showing the infinite possibilities for internal activities suspended in space yet vitally interconnected.

Glass Link The Lintas Bridge

Ian Ritchie, RFR

A bridge can be sculpture, a dynamic art form in its own right, a witty expression of absurdity, decorative furniture, a caricature of technology held together by steel, tension cables, and glass; a crossing point over a waterway, a roadway, a railway or just a space...

Up on the fourth floor of a heavy piece of architecture on the Quai de la Magisserie in Paris is a perfectly normal two-floor high internal courtyard, the plainness of its architecture accentuated by the dark grey of its painted walls. Looking into this "blackwall" of architecture are young advertising executives of the Lintas Group, dressed in bright colours going about their business. Creative

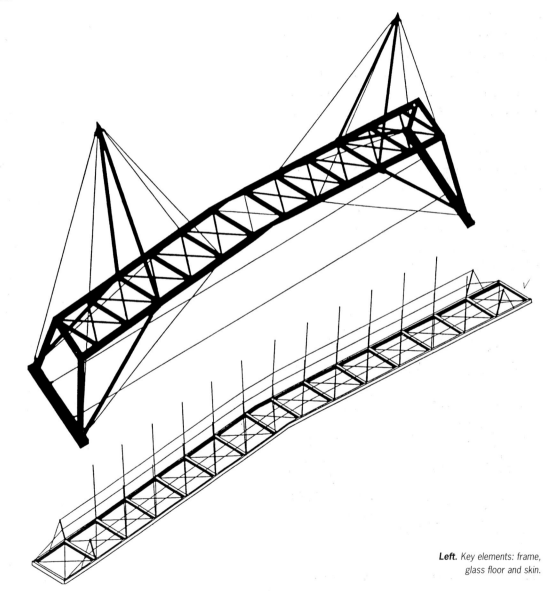

Left. *Key elements: frame, glass floor and skin.*

personnel rush about clutching portfolios, chattering in a variety of languages. Some of them stop to step out into the courtyard space and onto the Lintas Bridge…

"Diagonally across the courtyard it lies — or rather hovers — above a surface casually planned as a garden… a bridge made entirely of glass."

Architects dream of the all-glass object, but are usually dissuaded by reminders about safety, stability or treachery of one kind or another. This bridge proves that these inhibitions are all in the mind. It consists of two substances: glass for the entire skin, including the walkway, and steel for the frame and fixing. It looks and feels quite stable and yet there is an underlying sense of insecurity, that at any moment the whole thing could shatter dramatically.

Design in the past has been at pains to reassure us that architecture should look like something solid. We place a certain morality on architectural objects having a solid sense of purpose, allied to a solid sense of presence. In other words architecture is slow witted and a conservative art. By that definition then this bridge is illegitimate and not a piece of architecture. Yet history reminds us that this has not always been the case.

Reinventing the Glass Chain

As far back as 1920 a "Glass Chain" was formed by a group of very talented German architects and artists, the chief protagonist being Bruno Tuat who exhorted his friends "to free themselves of the tyranny of solidity". His call was responded to by Carl Krayl and Paul Scheerbart who each tried to outdo one another in the fragility and sheer explosiveness of their "glass architecture" projects. Plate glass in particular became the dream material of the

Right. *Exploded view of the components.*

Opposite page. *Inside the bridge during the day.*

20th century along with light steel lattice structures that have more in common with the spider's web or basket weave than the post and beam world we live in today.

In the hands of two generations of brilliant engineers, dominated by the British, we have come to expect designs in tension steel structures that are aesthetically beautiful as well as being wild with creativity. Perhaps it is something to do with the "boffin" tradition that takes pride in stretching techniques and making the "just possible" happen, rather than the easily possible occur.

It is no surprise to find that the designers of the Lintas Bridge—Rice Francis Ritchie—are all British or Irish, and once again we are being encouraged to explore their creativity in the city which enabled one of them, the late lamented Peter Rice to become a folk hero in architectural circles with his structural inventions on the Centre Pompidou.

Even at this small scale, we are struck by the audacity of the bridge. A horizontal shaft of glass, lit by a zany squiggle of coloured neon light running the length of the bridge. Occasional lines of wire remind us of the act of holding onto something, but there is no need for a handrail on the bridge as thin slivers of dense glass grip your feet along the walkway. Closer inspection of the bridge reveals that there are also devices for heating and ventilating the enclosed glass space.

Two steel cages rise up out of the courtyard to form the supporting structure from which the bridge is hung. The web-like quality of the support structure is visible all around. The glass panels resemble fish scales rather than windows. The cable masts herald a fair or a circus or the imminent arrival of the ringmaster. It is anarchic, yet here we are in a respectable and conservative heartland of Paris. If reassurance is needed we can always look out of the window over the roof tops to see the solid mass of the Left Bank, where a solid conservative world prevails.

What makes the bridge important rather than merely interesting is its tautness of manner, its very economy of means, the way the pieces of glass are clipped into place, the neon tube held by wires; in fact the whole thing suspended in a relaxed and reassuring style.

Philosophically it gives just the jolt that is needed in the field of architecture and design today.

As the creative minds of the young advertising executives trip back and forth over the bridge, one wonders whether they realise architecture can, once in a while, be as audacious and as dream-like as the world they work in?

Based on an article by Peter Cook in *Blueprint*, May 1987

A Walk Through History  The Ecology Gallery of the Natural History Museum

Ian Ritchie Architects

No one can fail to appreciate the beauty, skill and wit of architect Waterhouse's Natural History Museum. Equally no one can fail to recognise the invested energy of the Victorians in this part of London and their desire to illuminate and inform visitors — whether national or foreign — of the treasures of the world. Today no one can deny the importance of informing the visitor of the immense wealth and beauty of planet earth and of the fragile balance of man's relationship with it. This is the essence of the exhibition gallery on ecology which we were invited to design.

The Age of the Interactive Dinosaur

Our design for the structure of the Ecology Exhibition Gallery has been sensitive to two key understandings:

- the importance of the ecology "messages" to be conveyed to the visitor and our role in providing the vehicle for them, or as the brief from the client stated "creating a charismatic structure"
- Waterhouse's architecture, his linear and symmetrical planning, his spaces and their sequences, and the witty details which decorate the steel armature of the building.

Our understanding of the setting for the structure of the new gallery within this grand building created pre-conceptual guidelines for our design. The new structure must not touch the fabric of the building with the exception of the floor. It had to be designed in such a way that the whole thing could be dismantled and removed in ten years' time. The visitor must be allowed to enjoy the spatial qualities of the gallery not only from floor level but also within the volumes of space created within the structure.

We thought we should highlight the details Waterhouse layered onto the building, which include those motifs difficult to see high up on the walls and column heads.

Our design must offer the opportunity to create an exciting unidirectional route and sequence of exhibition spaces using contemporary presentation methods. In my view the design is in the spirit of the original intent of Sir Richard Owen who commissioned the Museum. The building was to bring to the general public, in a modern way, the latest understanding of the natural world.

Waterhouse used the spaces between the central columns and the external walls by placing large glass cabinets

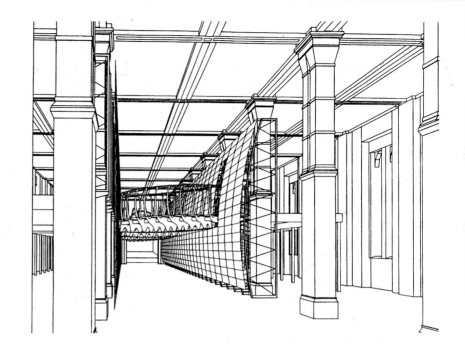

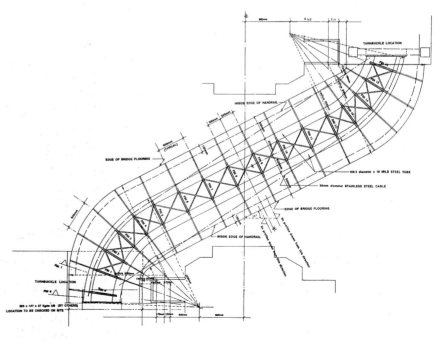

Above right.
Perspective from entrance.

Right. *Setting out: plan of first bridge.*

perpendicular to them. In our design approach we have restated the glass cabinet idea, but parallel to the external wall to create large "glass cabinets" into which both the visitor and the exhibit are invited.

The exhibition envelope is a fragile and delicate hands-off conversation with Waterhouse's architecture, in the spirit of ecology and modernity.

The bridges are a central feature of the design. We have used them to connect, to educate, to inform and to transport the visitor through and around the vistas and spaces created in the new gallery.

Bridges and the Quadrasphere

The visitor to the gallery will be moved by the spectacular back-lit glass wall towards the end of the room and the didactic walkways leaping across the void, with people appearing and disappearing through glass with seemingly no beginning or end. The asymmetry of the arterial space created by the curved and straight glass walls has been designed to evoke the notion that matter is in a perpetual state of dynamic equilibrium. The bridges threading their way in and out of the walls provide the stability to this moving world.

As the visitor begins to walk down the gallery they will be drawn into the exhibition through a layered glass wall on the right, which is curved and has the feel of a glacier. The ice wall effect is created by the use of "Optiwhite" sandblasted glass, lit from behind by cool coloured lamps.

On the left is a straight glass wall, which is glowing and flickering, lit by warm coloured lamps giving the illusion of heat and fire. The base of the wall undulates and there is a flickering glow alluding to lava coming from the bowels of the earth. Dark rubber on the floor evokes the earth and is made from recycled tyres. The visitor's attention will be diverted by an overhead bridge made of treated glass, set within an

essence the bridges represent the evolution of man's manipulation of the planet's natural resources.

Entering through an archway the visitor begins a more intense exploration of the exhibition and the huge Quadrasphere which creates spherical images relating to water and the oceans.

Moving behind the Quadrasphere onto the second "island exhibition" further up an inclined walk, the first bridge is crossed. The crossing gives spectacular views of the entrance gallery. The bridge deck is again covered with recycled rubber, referring to the

organic structural form, which is trying to register the fragile balance of ecology.

Etched into the surface of the glass walls are quotations relating to the theme of the exhibition. Waterhouse's terracotta fossil reliefs decorate the clear glass panels along the way.

Further along the arterial route the visitor will pass under three more bridges flying across the room. The purpose of the bridges will only become clear after walking across each one in turn, as they direct the visitor to each part of the exhibition at the upper level. In

earth.

The second bridge has a surface of wood, which is one of the primary materials used by man for building and for fuel. The visitor crosses the third bridge whose surface is metal, a material manufactured and processed from rock, and suggests another pause on the journey.

The visitor crosses the central glass walled space for the last time on a bridge where the panels are symbolically decorated to suggest hope and optimism but with a reminder of the precarious balance of nature and the evolutionary cycle. The glass floor is decorated with *Ginkgo biloba* and chestnut leaves, the first being a simple leaf shape and the sole survivor of a genus hundreds of millions of years old; the second is a complex bilateral shaped leaf of more recent times.

All the bridges are supported on a pair of braced hollow tubes 165mm in diameter, that sleeve through the rib-like fins of the walkway supports, whose ends turn up like antlers to form the posts for the handrail. Two stainless steel 30mm cables run under the structure acting as external prestressing tendons.

The final part of the exhibition is about the future, both its concerns and its optimism.

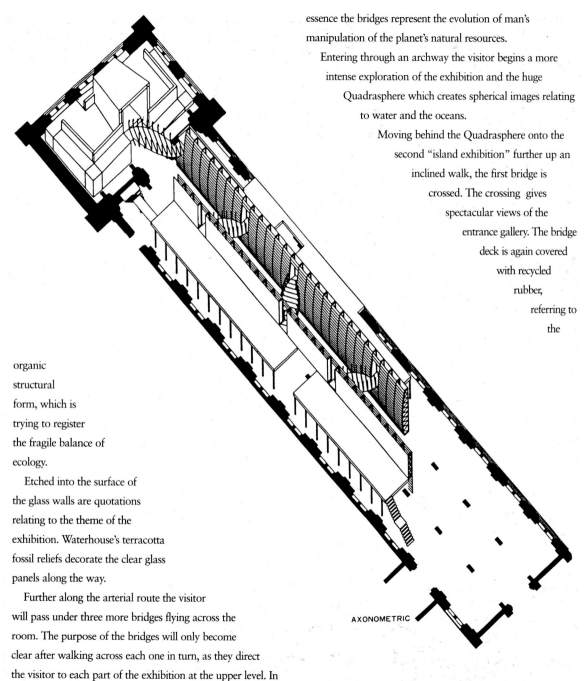

AXONOMETRIC

Left. Axonometric view of the Ecology Gallery.

Opposite page. The first bridge, with Quadrasphere in the background.

The Invisible Bridge in the Science Museum

Chris Wilkinson, Architect; Bryn Bird, Whitby and Bird, Engineer

This bridge is the central feature in the new *"Challenge of Materials"* gallery in the Science Museum. In addition to demonstrating the properties of steel and glass, the bridge challenges all previous structures in the goal to achieve lightness and invisibility in the interests of science and art.

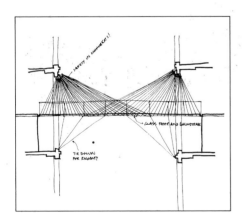

Far left. Concept section: bridging the void.

Left. Ken Unsworth: Suspended Stone Circle II, 1984. 103 river stones and wire.

The Spider and Postcard Theory

Our design intent was to create the lightest of bridge structures in order to demonstrate the "Challenge of Materials" in the concept. Early sketches were inspired from a modern version of the rope bridge held in a web of fine cables and a postcard sent by a friend in Australia. The postcard was a picture of Ken Unsworth's sculpture "Suspended Stone Circle II" showing a large mass of boulders attached to a web of fine cables that are raining down from the walls, like rays of sunshine, beaming in from an open window.

The architect presented the engineer with a picture of a spider's web and the postcard of the structure to symbolise the basic elements of this structure. The engineer responded by modelling structural ideas taken from the wires and stops of a grand piano. We came to the conclusion that we both wanted a transparent structure with a glass walkway, suspended across the gallery void, from a mesh of fine cable strands.

The bridge deck is constructed from individual glass planks 1.8m long, 75mm deep and 19mm wide, seated on two stainless steel bearers. The glass panels are laminated together to form 1.8m long panels that sit within the balustrade framework. The balustrade is made up of flat stainless steel

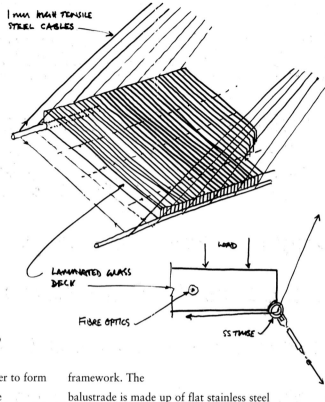

Above. Cables are stressed by "piano keys."

Top right. Elements of the suspended deck.

Right. Elements of the cable anchorage.

Far right. The bridge is supported by a mass of fine cables.

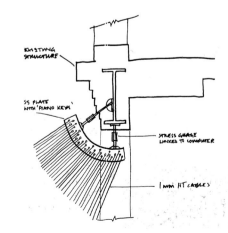

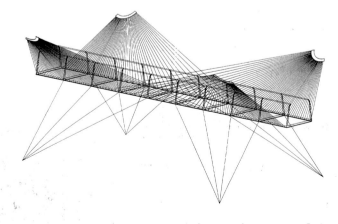

sections bolted together to form U-shaped elements fixed at 1.8m intervals across the walkway, supported on the two bearers.

The balustrades are infilled with laminated glass sheets to provide a 1.1m high parapet of solid glass protection, along the walkway.

From the four connection points to the main building, a web of 1mm diameter cables fan out to support the walkway at centres varying from 75mm at the ends, to 300mm over the mid span section. The cables pass around special pegs connected to the underside of the glass walkway. At the connection points to the main building, the cables are connected onto fan shaped anchor plates with "piano keys" for adjusting the tension in each cable. The anchor plate is connected to the building via "load cells" linked to a computer, which will relay the changing load patterns in the cables. Lateral stability is provided by four sets of cables attached to the main building, from below the bridge deck. Information from the strain gauges and other sensors will be connected to light and sound systems which will be activated by the loading and movement. The bridge will hum and glow in response to people using it.

Bridges are highly symbolic structures which throughout history have served to link communities separated by landscape divides. Successive developments in the technology of materials and construction have been reflected in design leading to lighter and more efficient structures. Lightness exists in nature and provides a goal to strive for. The Science Museum bridge challenges all previous structures in this goal to achieve lightness and invisibility in the interests of science and art.

The Garden Bridge

Cezary Bednarski, Architect

Green Gateways in the City

Bridges are amongst the most poetic and demanding structures. They are an explicit manifestation of human desire. They manifest the ancient human longing for being "there", over an abyss, a river, a torrent of cars, or the pounding wheels of a train. The only structure with a similar depth of atavistic meaning is a temple. Both are the most expressive points of reference for urban and non urban locations.

Fruit Trees Over Croydon

The windswept, sharp edged, rectilinear, shiny, reflective and humourless environment of Croydon provides a challenge for a new design idea. One that would create a strong place of "being" rather than a route with an anonymous identity; a crossing that would provide relief from the caustic reality of concrete, cars and tarmac; a bridge to reinstate a degree of humanity and greenery into a harsh environment.

The bridge offers the opportunity to create a green gateway between the two sides of a busy thoroughfare. The project mission would be fulfilled if people could be encouraged to take their dogs for a walk on the bridge.

The inspiration for this design concept can be traced back to the wondrous hanging gardens of Semiramida in Babylon, as well as to an urban bridge concept by Yuri Borzov. He won the 1987 Japan Architect bridge competition with a "countryside bridge" crossing over a Manhattan like cityscape. This was the catalyst for further research.

But it was the invention of a soil stabilising process patented

Left. Croydon Earth Bridge Concept (linking the Queens Gardens and the Fairfield Halls).

Below. Elevation.

The inspiration for this design concept can be traced back to the wondrous hanging gardens of Semiramida in Babylon

by Jack Blackburn that made the design possible and gave it intellectual and practical weight.

The bridge generates its strength and stability from the stabilised soil contained within a lightweight shell structure. The soil filled structure forms a flat arch spanning 23m. The low compressive strength of the stabilised soil — between 0.2 to 0.5 N/sq mm — requires a large cross sectional area to transfer the load. The 7m wide earth bridge is considerably wider than most pedestrian footbridges.

The large width of the bridge allows the planting of trees along the axis of the deck and rooting of climbing shrubs on the edges. The twin shallow arches of the bridge make the crossing a gentle and leisurely stroll rather than the climb that would have resulted if a single arch had been conceived.

But shallow arches carry a structural penalty — substantial horizontal thrust forces that have to be resisted by the abutments at each end. Consequently large sculptural buttresses of concrete at each end of the arch, conceived as rock outcrops, resist the thrust. The inspiration here has been the Landscape Arch natural rock bridge in the National Arch Park, Utah.

A central support mid way along the arch is made up of a lightweight tubular steel structure, emphasising the lightness of the earth bridge compared to a more traditional choice of construction.

Below. Concept cross section.

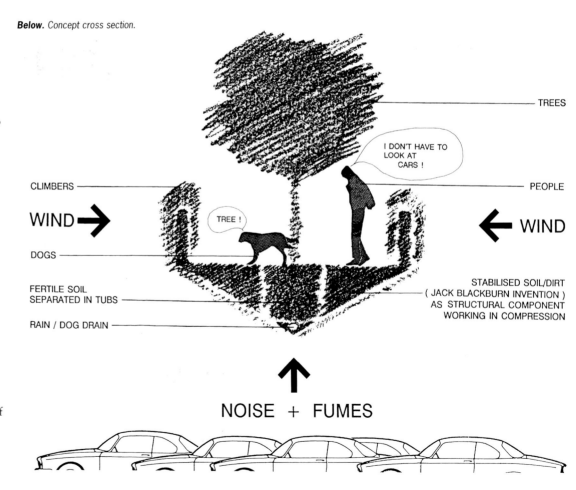

refurbished&recycled**bridges**

part *four*

Beaune Bridge Widening Cable Stays Save the Day

Michel Placidi, Engineer, with Alain Spielmann, Architect

Sometimes we put good ideas to work in the wrong place, sometimes we have good ideas that fail because we have used them with the wrong technique, sometimes we have a good idea that works and we can't explain its special qualities because it looks so obvious.

Right. *Detail of the new bridge cornice and handrail.*
Photo: Grant Smith

Cheers to Burgundy

We are in Beaune, in the Burgundy region, along the A6 Highway at the point where a minor road crosses over the A6. The problem we faced, as a design team, was what to do with the overbridge while dealing with the task of widening this stretch of the A6. It was critical that we kept the A6 open during the widening programme.

The obvious solution was to demolish the bridge and rebuild it on new supports which would be clear of the extended carriageway on each side. The Highway Company had done this before for other bridges along the widened route. Demolishing and bulldozing the structure would be quick and expedient, but the new build cost, the disruption to the traffic flow and the time involved would not be.

The overbridge at Beaune had four spans, the two end spans were 10m and two central ones were 20m. The trapezoidal end piers of the bridge run close to the verges of the carriageway. If we were to keep the bridge deck and the central pier, we would have to refurbish the bridge knowing that the two end piers would have to go, to allow for the road widening.

The solution settled on was to replace the end piers, picking up the support load using cable stays above the deck. In this way the integrity of the overbridge deck could be maintained with the minimum of disruption to the road below. The cost of rebuilding the bridge would also be reduced.

In providing tension supports for the deck, rather than a compression support, we had to ensure that the setting of the cable stays would not cause a reversal of stresses in the bridge deck. The dimension and position of the cable pylon over the central pier had to be squeezed into a tight space

and this required careful analysis. The external forces on the deck from the cable stays at the support points was balanced by transverse cables across the bridge deck, which also anchor the precast concrete cable block externally to the deck.

Steel H piles were driven on each end of the central pier for the pylon foundations. The pylon towers were formed using precast concrete segments, which culminate in a top section housing the cable anchorage.

The cable stays were covered in a black coated sheathing to contrast with the grey of the precast towers. The strident cerise colour of the bridge handrails, the cornices, the axis of the pylon and on the abutment walls, celebrates the wine growing region of Burgundy.

The Razel Firm has put a copyright on this technique.

Above. General view of the completed structure.

Right. Perspective view of cable attachment lugs to the bridge deck.

Far right. The original structure before it was widened.

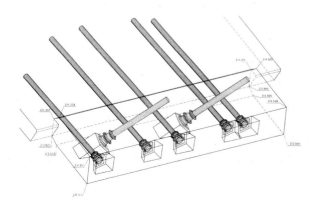

The Bridge Bank New Bridges for Old, or Old Bridges for New

David Ramsay, Engineer

We have been having informal discussions with a number of local authorities about the possibility of setting up a bridge bank in the UK. The idea has worked in Holland where there is a ready network of canals offering a huge number of similar short span bridges. Anyone wanting to sell a reconditioned bridge or those who have one to get rid of can contact the "bridge bank" which will attempt to match them up to clients who have need of a bridge for short term or long term use. It saves time and money on having to build a new bridge from scratch.

The Idea

A bridge which has been used for years can become superfluous and also lose its value. However it does not mean that such a bridge is worn out nor that it is only fit for the scrap yard. If a new home can be found for the superfluous one and there is a match between between the buyer and seller, then the old bridge will have a new life.

"The success of reuse is not always determined by the dimension of the bridge. An old lift bridge over a canal can be used as a light road bridge in a residential area" argues Peter Van Leeuwen of DHV Dutch office, the first company to set up the idea of the bridge bank. "Clients looking for a bridge to fill a particular need get in touch with us" he adds "If we can find them a second hand bridge that fits the bill, it will be cheaper."

The customers supply details of the type of bridge they need, giving the span, width, type, and other specifications.

The information is then fed into a data base and checked to see whether a match can be made.

"We put the potential customer in touch with the seller and they set up a site visit to see the bridge. We act like an estate agent brokering the sale and taking a fee if the sale goes through. We have completed a few sales in the past year" confirms Van Leeuwen. "As an example of the principle, a 922m fixed bridge was given to Bangladesh as a gift of the Dutch government. It had been part of a 2785m bridge forming temporary access to an island during the construction of a sea storm surge barrier in the Netherlands' Eastern Scheldt. It now crosses the Karnapuli River near Chittagong in Bangladesh."

The Reality

The idea has the potential to work well in Holland with its large network of canals and huge reserve of similar bridge spans. Naturally there is a big list of sellers on the bridge bank data

Left. *Bridge moved from Opsterland to Gramsbergen.*

Right. *The bridge at IJsselstein originated in Apeldoorn.*

Above. *The Karnapuli Bridge in Bangladesh was originally used in the Netherlands.*

base, but many of them don't make an exact fit either because of the location or the crossing proposed by the prospective buyers. A successful match is often a compromise with a buyer having to adjust or adapt the site to suit the dimensions of the recycled bridge.

Bridges become available because of projects to widen roads or navigation canals or because they are simply not strong enough or too old to carry increasing traffic loads. Such structures can be successfully adapted for more lightly trafficked roads.

The possibilities for re-use are numerous. For example a used bridge can be placed where a temporary river bank connection or a road crossing is needed for traffic diversion. A narrow bridge can be replaced by a wider one or a single carriage way can be doubled with a parallel bridge. A lift bridge rather than a fixed bridge might be better as a

replacement structure, for aesthetic reasons, for bridges which lead to and from residential areas.

Due to lack of funding a local authority may be forced to postpone the building of urgently needed bridges for many years. Using the bridge bank can drastically lower the cost of a bridge undertaking and the design fees and so help to stretch a limited budget a lot further.

Dutch bridges tend to be shorter and in many ways similar in construction and relatively modern. The bridge stock is under 50 years old and designed in such a way as to make dismantling and transportation relatively easy.

UK bridges on the other hand are older and often replaced when they have very little or no useful life left. Many short span bridges are masonry constructions which are difficult to relocate. But despite that, there are opportunities for bridges built in steel or timber that can be recycled for pedestrian and

road bridges.

Although the UK may not have extensive canal networks, we find that a lot of road structures have common spans which are repeated in different regions around the country. The idea will work best in areas where clusters of local authorities, such as those on the outskirts of, say, London or Birmingham, can network together through the bridge bank and not have to transport them far afield.

Response to the bridge bank has been good. There is a lot of interest, even though people are not sure how the bridge bank will work. The first move has been made. A data base has been started. The logging of bridges for sale will take some time to capture as information from all over the UK is being compiled. But once it has been set up then the bridge bank net will be open for enquiries.

Broxbourne Bridge Study

Ronald Yee, Architect

"The search for alternatives is fundamental to creative design. There is a need to go beyond the known, the obvious and the satisfactory. Being positive is a choice. We can choose to look at things in a positive way. We can choose to focus on those aspects of a situation that are positive. We can search for benefits."

Edward De Bono, "Six Thinking Hats".

The Structural Problem

Our brief was to refurbish, upgrade and aesthetically improve a collection of circa 1900 bridges, comprising three brick arched viaducts and two steel girder bridge spans that carry the B194 over a rail line and a branch of the River Lea at Broxbourne, Hertfordshire.

Broxbourne Bridge is of strategic importance to Nazeing as it is the only road connecting Nazeing to Broxbourne. It is also the access road for Broxbourne BR Station a main line stop on the busy London to Cambridge line and Stansted Airport and the access to the hamlet of Nazeingbury.

The bridge is located in the heartland of the Lea Valley Recreation Park and close to the Broxbourne Lido Centre. It will also become the principal access for a future wildlife reserve, golf course and leisure park which will extend from Broxbourne to Hoddesdon, when gravel pit extraction in the area runs down in a few years' time.

Waterscape, trees, parkland and wildlife conservation characterise the eastern approaches and the Essex half of the bridge. The built fringes of Broxbourne Town, the county boundary of Hertfordshire, a public house and the historic new river canal crossing characterise the other side.

The bridge is in the joint ownership of Hertfordshire County Council and Railtrack. The Council is responsible for maintaining the brick approach viaducts and the roadway surface. Railtrack is responsible for the steel composite bridge spans over the railway and the River Lea navigation waterway.

Initial inspection of the brick arches suggested that the structure was in sound condition requiring only localised remedial work and a good clean. The approach viaducts have been assessed by the client and deemed to comply with the new Euro standards which will permit 40t vehicles over bridges.

Our preliminary studies indicated that the composite steel spans required repair and strengthening. We felt that the refurbishment work could be carried out in situ with one lane closure necessary to maintain traffic flow. The steel girder beam had been strengthened in the late 1950s and jacked up over the railway span to create sufficient clearance for electrification of the line.

The Traffic Problem

Broxbourne Bridge is not a pretty place to be between 7.30 and 9 am, and 5.30 and 6.30 pm every weekday. It is a major traffic bottleneck at peak times and a serious hazard for pedestrians using it to access the Lido Centre, the waterside amenities of the parkland, river and navigation canal. Traffic trying to turn right into Broxbourne Station approaching the bridge from Nazeing is hampered by through traffic from Broxbourne. As a result vehicles back up over the bridge and beyond. Cars attempting to turn out of the station onto the B194 are blocked by through traffic from both directions and the vehicles queuing on the bridge trying to turn into the station.

To compound the problem there is a marked kink in the bridge alignment between the rail span and river span creating a blind

Top. *Existing steel girder and brick "jack arched" spans in need of repair.*

Above. *Existing brick arched approach viaducts still in good condition.*

Broxbourne Bridge is not a pretty place to be between 7.30 and 9 am, and 5.30 and 6.30 pm every weekday

bend, with an access road leading onto the bridge a short distance away. The bridge along this section is dangerous to pedestrians as the bend is too acute for large commercial vehicles causing them to infringe onto the oncoming lane or to mount the pavement! The number of near misses to pedestrians is legion. The scattered fragments of sheared off wing mirrors lying on the bridge tell their own story.

Old Nazeing Road joins the bridge just beyond the blind bend and meets it at the high point, supported on a separate brick arch viaduct. Clearly Old Nazeing Road is too narrow to allow two way traffic onto the bridge. If a vehicle has managed to turn off the main bridge onto the viaduct of Old Nazeing Road, a vehicle approaching the turn in the other direction has to reverse 120m to a passing point.

The traffic problems can be summarised thus:

(i) Station approach: during peak flow, congestion is caused by vehicles wishing to turn across oncoming traffic, into the station, backlogging traffic over the bridge. In addition, vehicles wishing to leave the station have difficulty in doing so because their exit is blocked by traffic turning in.

(ii) B194 and main bridge: due to the acute bend between the bridges over the railway and the river crossing, there is the danger of commercial vehicles mounting the footway or scraping up against the bridge parapet walls.

(iii) Junction with Old Nazeing Road: Old Nazeing Road is too narrow for two passing vehicles. Problems are caused when a vehicle wishing to turn off the B194 onto Old Nazeing Road, a blind turn, is confronted by an approaching vehicle. Either the vehicle mounts the kerb, creating a hazard to pedestrians, or reverses 120m to a passing point.

Bridge Strengthening

Top. Main bridge unable to meet new EC 40t loading requirement and needs strengthening.

Bottom. Existing structure already at limit of railway clearance envelope and any new works must be above soffit level.

The Problems

Top. Current road layout causes problems during peak time traffic.

Middle. Creation of a third lane on a wider bridge helps through traffic but not station exit.

Bottom. A mini roundabout gives order to traffic flow and has a desirable calming effect on traffic.

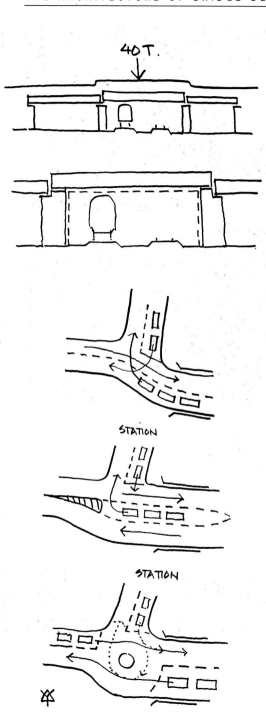

A Solution

Our initial thought was to build two new brick arch spans over the rail line and over the river to smooth out the kink in the roadway over the bridge and to incorporate a wider carriageway. However the need to keep the road open for traffic at all times, plus the cost penalties for rail closures and the limited budget, effectively ruled this option out.

We then went on to develop a design which removed the pavement from within the existing carriageway and provided new walkways on the outer face of the structure, external to the bridge parapet walls. This effectively widened the roadway to 3.5m and at the same time allowed refurbishment of the bridge and carriageway, a lane at a time, without a closure.

Removal of the footway from the existing deck not only achieved the desired separation between pedestrians and road traffic, but also gave an extra 1.75m to the carriageways.

Strengthening of the composite steel span is achieved by laying down a new reinforced concrete slab over the top of the jack arches in two halves, allowing a single lane to be kept open at all times. To lessen the amount of fill required for a smooth roadway elevation, we suggested raising the third span from the Broxbourne end to match the level of the second span and adjusting the approach span and fourth span. This realignment is required aesthetically to

Top left. *New mini roundabout at station exit.*

Top right. *New walkways placed outside of existing bridge structure.*

Below. *The new walkways visually unify the different structures.*

give a constant soffit line.

Proposed traffic improvements include the provision of a mini roundabout to the station approach. The viaduct approach from Old Nazeing Road is made one way with no entry from the B194 and with no right turn from Old Nazeing Road onto the B194.

Aesthetic Emphasis

The new footway structure running the entire length of the parapet wall masks the external line of the bridge, unifying the differing structural forms rather like the elevated topiary in an ornamental garden or a tree lined avenue in a street of stark terraced housing.

It is a composition that is structurally efficient, yet it captures the character of greenery and countryside and visually transforms the overbearing mass and dark forms of the main bridge structure.

Where the footway structure is able to be taken to the ground, the new footway is supported on structural "trees" that are fabricated from recycled steel lamp posts.

Over the railway, where the structure cannot be taken to the ground, the footways are supported on "outcrop branches" of steel, bolted to the steel girders.

The walkway deck is fabricated from galvanised checkerplate that is carborundum coated on its wearing surface. The parapet guarding is formed from interwoven light and dark green metal strips that are gradually varied in width so that the parapet is unobtrusive on the viaduct approaches but solid over the railway, thus conforming with railway regulations. The parapet is topped by a well-seasoned oak handrail.

The aesthetic appreciation of a bridge is dependent to some extent on its immediate surroundings. To upgrade the appeal of the refurbished Broxbourne Bridge we must also consider its immediate setting; the bare concrete and blistered hardstanding

of the Lido car park area. We have proposed landscaping to the leisure centre car park area, creating a riverside walk and picnic area to match the amenity available in the other areas surrounding the Lido.

Pedestrian lighting is provided by semi-domed recessed wall fittings fixed to the outer face of the existing brick parapet.

Total costs for the proposed works, additional walkways, strengthening to composite spans and resurfacing is half the cost of a conventional new bridge deck and many times less than the price of a new bridge. It represents value for money and a long term solution to the traffic and pedestrian problems of Broxbourne.

We hope its architecture adds to the quality of the environment and will be enjoyed by the users, should it ever be built.

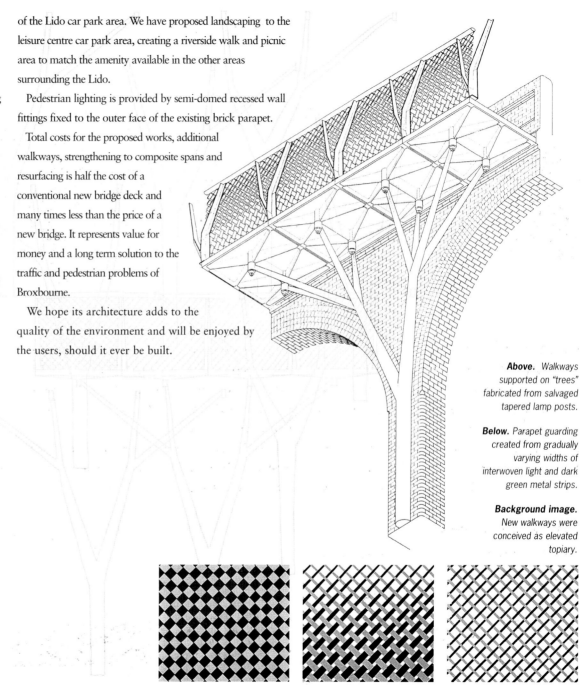

Above. Walkways supported on "trees" fabricated from salvaged tapered lamp posts.

Below. Parapet guarding created from gradually varying widths of interwoven light and dark green metal strips.

Background image. New walkways were conceived as elevated topiary.

The Big Slide

Marsh Mills Viaduct

Frank Rowley, Engineer

*"Oh winding bridge
in sorry state*
how can you be replaced,
yet still avoid that awful fate
of making travellers late.

Just think: a skater seems
to glide
with easy strength and
grace,
can it be a simple slide
will put you back in place ?

the answer's yes, the
prize is great
it pays to risk and

innovate."

Frank Rowley, "ode to the big slide", 1995.

Minimising Lane Closures

It is interesting to see how priorities on construction cost and in particular the penalty for lane closures can bring about innovative ideas in bridge refurbishment schemes. If the penalties for lane closures were a lot less onerous, it is likely that we would put together a humdrum design proposal for Marsh Mills, which would not get a mention in this book.

Minimising lane closures was the challenge given to the design and build team when assessing the options for refurbishing the Marsh Mills Viaduct. The client requirement for minimum traffic disruption was enforced by setting down a series of heavy penalty charges which depended on the time of day and the number of lanes closed. We reckoned the refurbishment contract bid would be won or lost on lane closures. Any construction method which required extended lanes closure would prove to be uncompetitive.

Both the contractor and engineer addressed this problem separately and pooled ideas to come up with a number of options including:

- complete closure and demolition of each viaduct and replacement on line

- part demolition of each viaduct, retaining one half for traffic flow while rebuilding the other half.

- provision of temporary viaducts alongside, followed by demolition and replacement of each viaduct on line.

The outcome of this brainstorming exercise, where we compared benefits and disadvantages of the options, led

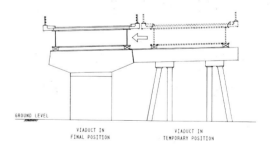

Top. Steel erection on slide track in temporary position.

Middle. Slide track.

Bottom. Arrangement of bridge deck in temporary and permanent positions.

was suffering from the effects of alkali-silica reaction, which is a break up of the cement properties of the concrete. Strengthening works were considered a temporary measure, and enough to extend the life of the viaducts until the new A38 flyover was built. The flyover was completed in 1992 and the client, the Department of Transport, immediately put in hand the demolition and replacement contract for the Marsh Mills approach viaducts.

Our first priority was to design a lightweight deck that would minimise the jacking force needed for the slide. The deck we designed comprised two steel plate girder beams 1950mm deep, 8.6m apart, spanning 35m to 53m between pier supports. The longitudinal girders were connected by cross girders 686mm deep at 3m intervals. The running surface of the carriageway was built up using precast concrete planks that span 3m between the cross girders. The precast planks were topped with in situ reinforced concrete to give a total depth of 230mm. Precast concrete parapets cantilever out from the main girder beam.

Time was critical in all operations so that is why we chose permanent participating formwork in the shape of precast panels. This meant that all deck construction work above ground could take place without the need to scramble below deck level to position and remove temporary formwork and scaffold material.

The first task on site was to install the temporary pier supports adjacent to the existing viaducts. The piers comprised two sets of four 600mm tubular steel driven piles, two raking and two vertical. The temporary piles were driven with embedment lengths of up to 20m below ground and a free standing length above ground varying from zero, rising to 15m near the flyover intersection. The temporary piles were bridged by a deep crosshead

us to design a new viaduct alongside the existing one and then slide it into position. Traffic could be diverted on to the new viaduct, even though it was built in its temporary position, while the existing viaduct was being demolished. Each new viaduct would be closed for only 48 hours, while it was being jacked sideways. The approach is well proven for overnight "possessions" in the railway industry. This proved to be the most economic solution and won us the design and build contract working with Hochtief.

Each viaduct was 400m long and had to be jacked

sideways a distance of 12m. This is one big slide by any stretch of the imagination.

Permanent and Temporary Structures

The new viaducts replace two existing viaducts leading off the A38 down to the Marsh Mills roundabout near Plympton. Problems with the original viaducts arose some years after construction, when sections of the concrete structure were deteriorating at an abnormally high rate. Further forensic engineering revealed that the structure

Top. *Merge viaduct before slide.*

Above. *Merge viaduct on completion of slide.*

beam and the sliding track which runs on it. The main girder beams of the deck rest on the sliding track. Altogether, 76 steel piles were driven for each viaduct.

While the permanent deck structure was being built off the temporary pier supports, the permanent pier supports were being built below the existing viaduct deck. As the depth of the new deck was much greater than the existing deck, there was sufficient headroom below the existing bridge deck to form the new piers. The alignment of the new pier and the temporary pier was critical. The maximum centreline deviation of 10mm had to be achieved while the levels of the slide track had to match exactly.

Demolition of the old viaducts could commence only once the traffic had been diverted on to the new slip road on its temporary supports. The demolition of each viaduct was a significant element of the work in its own right, taking three months to complete. Altogether, well over 20,000t of concrete was broken out and removed from the site.

The Big Slide

To slide the bridge deck from the temporary to the permanent pier, all the slide tracks on the piers had to be parallel to each other. The level and alignment had to be set precisely to match the transverse slope of the permanent pier. This careful planning ensured that there was no twist in the span length of the deck.

When loads are being moved on slide tracks there are two levels of resistance forces the jack has to overcome—the static friction and the sliding friction resistance. The static friction load has been estimated as 5% of the weight of the deck structure. The sliding friction is usually about half the value of static friction, once the deck is moving. One of the main problems with friction is that it can never be predicted with certainty. It is therefore prudent to design the sliding system to absorb a maximum value which past experience has shown will not be exceeded. At Marsh Mills all the equipment was designed assuming that static friction would be 8% of the vertical load, which was 440t for the merge viaduct section. It was expected that the static friction would be nearer 5% of the vertical load and sliding friction around 2.5%. Because the merge viaduct was on a down slope of 2.85%, the effective sliding friction was approximately zero. This meant that the brakes were needed and these were active during the slide.

The moving forces were applied by jacks and pulling rods positioned at six out of the nine pier locations. The hydraulic jacks were located at the ends of the permanent piers, opposite the ends of the temporary piers. The new deck was pulled into position by high tensile steel bars anchored to the deck and fed through the hydraulic jacks. The force applied to each jack to move the bridge, was 34t. The jacking force was maintained by a central control which balances the hydraulic pressure acting on all six jacks.

The continuous slide tracks were made from polished stainless steel plate. Special lubricants were applied to the steel plates before the bridge deck was built. These measures, together with essential cleaning and drying of the track during the slide, contributed to the low friction levels.

Moving large loads with high levels of accuracy is usually done at very slow speeds. The maximum rate of slide, which includes retracting the hydraulic jack after each pulling stroke, is about 5m per hour. This means a minimum time to move each viaduct was 2.5 hours, although we made an allowance of 8 hours to allow for bad weather and other contingencies.

The Millau Viaduct

an Introduction

It is two years since I was first shown the concept sketches for the bridge, but what a bridge! The piers were to be higher than the Eiffel Tower and in length the bridge was twice as long as the Champs Elysées. The bridge spans the gap between two massive limestone ridges separated by the Gorge du Tarn, on the beautiful but wild landscape of the South Central Massif, near the medieval town of Millau.

As I sat in Alain Spielmann's office sipping an espresso and talking over his ideas in design, he confessed that he was worried. The Millau project was the biggest he had ever tackled, it was going to take considerable energy to get the design right in his mind, and there was no guarantee of a design fee at the end of it. Should he have taken the plunge or continued with the scale of work he was used to? It was such a monumental task for a small practice and one which could so easily inflict a fearful bashing on staff morale and self esteem if it came to nothing. Would he and the team get over the disappointment quickly? But opportunities like this arrive once in a lifetime—the challenge was too irresistible to ignore.

Alain Spielmann, along with thirty other creative designers, one of whom was Foster and Partners, were invited by SETRA—the French highway and bridge authority—to comment on a number of standard designs developed by their in-house engineers for Millau. It was the first stage of a two stage concept development which was to take two years to finalise. The word was that the project was needed fairly urgently as the motorways on each side of the gap were going to be under construction in a year's time.

The thirty designers were whittled down to five, each was paid a decent fee to develop a particular design solution that SETRA had chosen for the final concept study. SETRA had also nominated the French engineers who would team up with each creative designer. Alain Spielmann was invited to develop the concept for a massive concrete arch with Jean Muller Engineers, not the cable stayed option he might have preferred. Foster and Partners was invited to develop the cable stayed option, another team a box girder bridge with splayed piers, another a steel truss bridge on twin supports and the fifth team a hybrid ribbon beam type bridge.

I have chosen the first stage concept designs for a cable stayed bridge developed by both Alain Spielmann and Foster and Partners and the exquisite winning solution that emerged after the second stage study by Fosters. It is a tragedy that there should be any losers in such creative and imaginative design.

David Bennett, September 1996.

the**viaduct**

part *five*

Le Grand Viaduc de Millau Limits of Creativity

Foster and Partners, Architect

"After a lunch of excellent figs, I took a guide and went to see the Pont du Gard. It was the first Roman construction that I had seen. I was expecting a monument worthy of its builders, but in fact it surpassed my expectations—the first time this has happened in my life. The skill of this simple and noble construction struck me much more since it stood in a desert of silence. Its solitude made the object even more striking and thus the admiration stronger—this so-called bridge was nothing more than an aqueduct."

Jean-Jacques Rousseau (1712-1778)

Reinforcing Connections

Rousseau's philosophy was founded on a deep respect for the natural world. Yet we can see from the quotation that even Rousseau was profoundly moved by the majesty of the Pont du Gard. This teaches us that the problems of scale and building man-made edifices in places of outstanding natural beauty can be resolved in a satisfactory way. It also shows us that a man-made result can be an object of inspiration for even the greatest of nature lovers.

For the Grand Viaduc de Millau it is important to begin by defining the design aims which are appropriate for such a bridge, in such a beautiful location in the South Central Massif at the head of the Gorges du Tarn. We defined the design aims only after careful analysis of the landscape, the town, its history and the perceptions of those who will see and use the viaduct.

Above. *Photomontage of the final stage cable stayed concept design, superimposed over the Tarn Gorge.*

Below. *The town of Millau with the ruins of a medieval bridge, plus a modern bridge in the foreground.*

Right. *Map of the Aveyron area showing the location of the Millau Bridge and the route of the A75.*

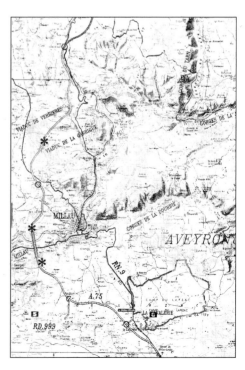

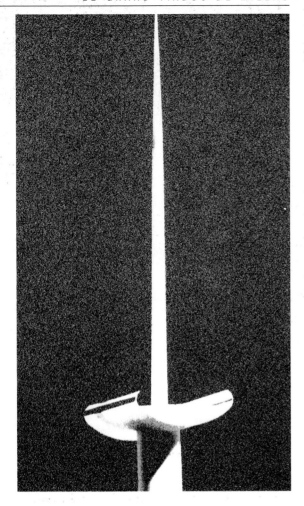

The bridge is 2.5km long, elevated some 240m above the valley. It is grand in scale by any definition. The proposed trajectory of the A75 Clermont Ferrand motorway bridges the gap between two limestone plateaux—the Causse Rouge to the north and the Causse du Larzac to the south. A canyon of sheer limestone cliffs overhang the Gorge, creating a stunning backdrop to the green valley floor and silver snaking river.

The alignment and scale of the bridge must ensure that the little town of Millau and the viaduct will be inextricably linked. The new viaduct will make a powerful contrast between the quiet of the landscape and the historic town of Millau clearly visible in the distance. The contrast must be handled with great sensitivity in order to minimise the impact of the bridge upon the town.

An approach that we favoured would be to minimise the physical impact of the bridge by reducing the points of contact that the new bridge makes with the valley floor. However if we look at the Pont du Gard, we see a massive structure at the limits of available technology of the time, which gives it its strength and creative energy. This is the idea that we tried to follow in developing the design concept.

The town's population has been on the decline for many years as the young citizens leave in search of better employment opportunities in the more northern regions or to the south, towards Provence and Rousillon. Millau nowadays relies heavily on tourism and selling cottage crafts. It no longer has an export trade since the halcyon days in Roman times, when it was famous for pottery. The clay and kaolin has long been exhausted. The chestnuts, oaks and maples that proliferated on the valley slopes and plateaux, had been so drastically logged that it has taken many hundred of years to re-establish them.

The town's tourist trade is largely due to its position at the head of the Tarn Gorge. At present the RN9 runs through the town to cross the river Tarn on route to Roquefort and Valence to the east and Toulouse to the south. This detour will not be necessary for those journeying through once the new viaduct is in place. The viaduct and new road crossing pose a possible threat to Millau's tourist trade. For this reason the connections between the bridge and the town must be reinforced.

This suggests additional design objectives for the viaduct:

- *that people crossing the viaduct should be aware that they are crossing the Gorges du Tarn. The crossing should be experienced as an event: perhaps the major*

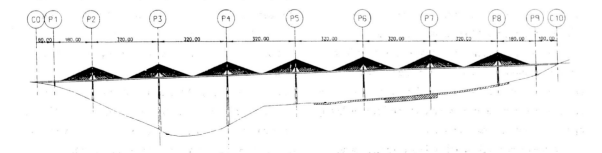

Above left. *Model of the cable mast and pier support to improve the aesthetic appearance.*

Left. *An equal span cable stayed bridge with a constant rhythm of spaces worked best.*

Above. *The bridge profile of the first stage concept design developed for SETRA.*

event on the road between Paris and Barcelona

- *that Millau should embrace the viaduct as a new symbol for the town's future, as the third millennium begins.*

History also reminds us that there has been a crossing over the Tarn in Millau since 1 AD, and that the first proper bridge was built around 1000 AD. It is fitting that as we approach the year 2000 Millau will have a second bridge.

Preliminary Aesthetic Appraisal

We examined the aesthetic design of various bridge solutions proposed by SETRA, the French highway authority. They had looked at a number of conventional bridge solutions and had set out their findings in a briefing document for us to comment upon. One solution was a box girder bridge, another an arch, and yet another a cable stay option, a beam and post and so on.

We came to the conclusion that the pier spacing of an economical beam and column type bridge should increase over the valley section and be as few as possible. The rhythm of the spacing should be reduced as the viaduct climbs out of the valley. However, for a cable stayed solution, a constant rhythm of pier spacing seemed to work best.

In our view the cable stayed option was preferred to a beam and column construction, because it had all the potential to be a magnificent and exciting structure. We remodelled the stay proportions, the bridge deck profile and pier supports of the original design SETRA presented to us and sent these suggestions back to the client, with our recommendations.

Many months passed by until we were invited by the client to work with a team of French engineers comprising Sogelerg, EEG, and Serf to further develop and explore the options of a multi-cable stayed viaduct for Millau, in line with our preliminary appraisal. We were one of five teams commissioned by SETRA to develop one of five bridge options for Millau.

The Second Stage Design Study

The final design study followed on from the preliminary appraisal, which had led us to determine the spacing and rhythm of the piers and the structural development of the cable stays. The final stage study refined this process and brought us on to a clear and powerful statement of the bridge solution.

Our thought process is summarised in the following statements, starting from a beam and post idea and developing the reasoning for the cable stayed option. We illustrated the design issues with a series of diagrams:

1 *"… the crossing is not only that of the river Tarn or of the Plateau of France but is also on the scale of the whole site from causse to causse …"*

The very wide opening of the site, which is perceived as a large space, suggests a very even distribution of the piers since no intermediate relief punctuates the crossing between the two causses.

2 *"… piers must be located evenly between the causses. If we try a 170m span the result is a cluttered look with too many piers …"*

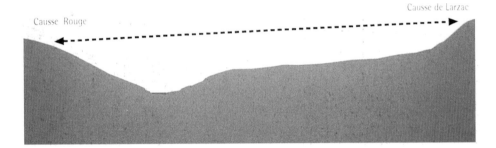

Causse Rouge   Causse de Larzac

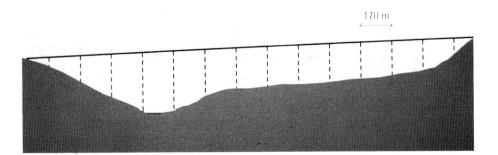

170 m

If we stick with economical spans of concrete and steel box girder construction, then a 170m span appears to be a good engineering solution. We are left with 14 piers which are difficult to locate along the valley slopes. The piers also cut the spaces between them into rectangles thus forming barriers in the landscape.

By pushing the spans out to their limit for box girder construction it is possible to increase the span to 260m. In this case the thickness of the bridge deck becomes very deep and will transmit large wind forces and bending stresses which are unacceptable to the slender piers.

The deck must be made thinner in order to reduce the wind forces, at the same time the span should remain at 260m or more. A cable stayed span will reduce the deck thickness and increase the span if required.

) "… the longer span increases the deck thickness, which attracts high wind forces and becomes difficult to support on slender piers …"

Cable stayed bridges are well suited to spans ranging from 250m to 500m without much increase in the bridge deck thickness.

) "… 350m is an optimal distribution for cable stayed spans. The location of the seven piers respects the site below and makes for an economical structural solution …"

Under the effect of the temperature variations, a bridge deck of 2.5km expands and contracts with a displacement at the ends of the structure that changes by as much as 1m.

) "… the expansion of the deck is significant due to its length. Piers must be made flexible to allow this movement to occur if they are to remain slender …"

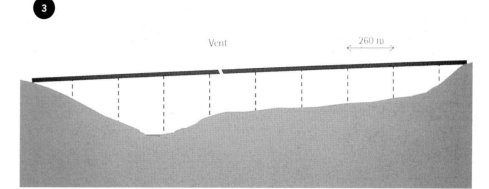

3

Vent   ← 260 m

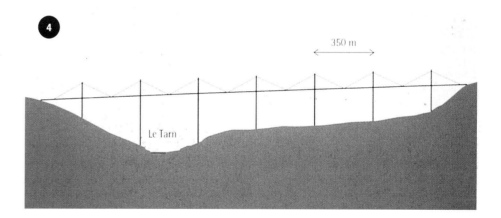

4

350 m

Le Tarn

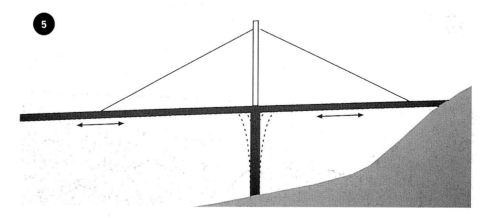

5

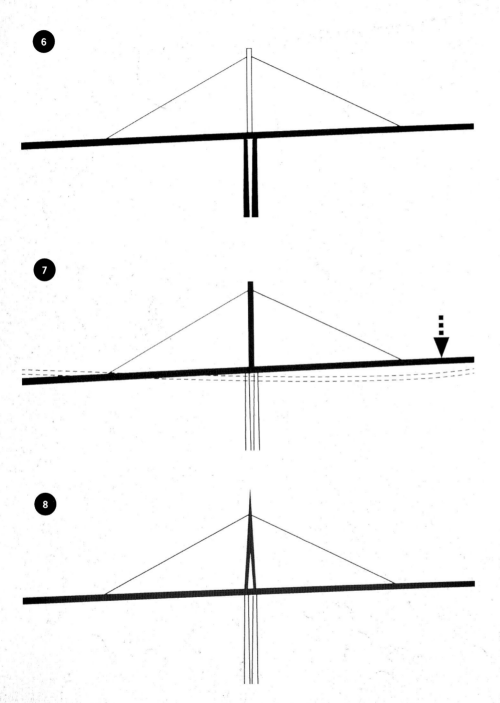

In order to limit the displacement to the piers, sliding bearings could be placed between the deck and pier. However a heavy concrete deck will generate high frictional forces which would cause high bending forces in the piers. It would be a good thing to make at least the end piers flexible in construction.

6 *"… solution: make the piers more flexible by dividing them into two thin shafts…"*

In order to guarantee the flexibility of the piers for longitudinal movement, the single shafted piers are replaced by split piers with the same overall dimension.

Conversely, the cable mast head must be made rigid in order to hold its shape and equilibrium when a span is subject to bending, pulling on the stays which are anchored to the mast head. The bending load between adjacent spans may not be identical due to traffic flows.

7 *"…problem: a slender mast is not rigid enough to retain the stays which take up the loads imposed on the deck …"*

In order to ensure that the staying system works properly, the mast head is made rigid by making it an A frame, with two downstand legs.

8 *"… solution: a mast head with two spaced apart legs, creates a more rigid and stable shape …"*

A harp shaped or parallel arrangement of the stays, which we preferred for aesthetic reasons in the preliminary design stages, was not the best choice for the structure. The decision to design a triangular mast has led to the grouping of anchors for the stays in the upper half of the mast.

"...Stays are best secured to the mast head in a fan arrangement, which is consistent with the triangular geometry of the mast head ..."

The curve on plan which we have deliberately given to the viaduct eliminates the unpleasant sensation the motorist may experience on the arrival of a steep slope at one end as it falls away in a straight line across a breathtaking deep valley. The curved alignment of the bridge also gives motorist a good link with the plateau and winding roads beyond.

"...The viaduct slopes steeply on account of the differing plateau levels at each end. A straight line across the bridge would give the motorist a 'slide' effect. A curved alignment on plan eliminates this feeling ..."

The curved alignment of the bridge gives motorists a lateral view of the viaduct which would not be available with a straight line crossing.

The Last Word

Once the final design is approved by the client we would recommend the construction of a large model, possibly 10m long. The model should be displayed against a backdrop of the real site in an Information Centre in Millau, which could serve as a museum to commemorate the construction of the viaduct.

Right. *A harp shaped arrangement of stays was not the best choice for the structure. A fan shaped arrangement worked best for the grouping of the anchors.*

Millau Viaduct
A Cable Stayed Solution
Alain Spielmann, Architect

In designing such a huge bridge we have very little experience by which to judge the scale and proportions. Designing a house or a building is quite different, we are disciplined by the proportions of the human, for scale, for enclosure, for space and for comfort. For Millau we have only the scale of the hills and the width of valley to measure the lines and the proportions of the structure.

Below. *Character of the houses hewn from local stone and built into the hillside is typical of the region.*

Bottom. *View from the top of the plateau looking back towards Millau near the site of the crossing.*

The Particular Point

There are many ways in which the character and environment of the bridge site can be explored. It is always necessary in my view to walk the hill side, the valley, the town and the river bank many times. Detailed examinations of photographs, maps, drawings and books have been invaluable in creating a starting point for initial design ideas. It was critical to almost loose myself walking in the hills around Millau, in the forests and across the limestone plateaux, in order to gain a better insight of the natural lie of the land, the contours, the rock formation, the shape of the valley, the isolated spaces between the sky and the river…seen from different angles, in different light and under varying cloud cover. On a misty morning for instance or when there is low cloud it was evident that most of the bridge would be shrouded in cloud. It is obvious once you realise it, but it could so easily be overlooked if that experience were not discovered.

The bridge is twice the length of the Champs Elysées and one and a half times longer than Varazzano Bridge in New York. Its height from cable mast to valley floor is 20ft taller than the Eiffel Tower and as high as the Canary Wharf Tower.

In the early morning sunrise and through the hottest part of the day, reflected light from a long ridge of limestone on the Causse Rouge dominates the hill tops overlooking Millau. When the Millau Viaduct is built, its silhouette will cross the skyline to the right of this ridge. Somehow we must find a way of balancing this change in perspective so that the presence of the bridge contrasts with the ridge. We should consider floodlighting the limestone ridge at night, creating a son et lumière when the bridge is lit up. It was important for me also to meet the people of Millau, the local farmers and woodmen to complete my local knowledge and understanding of the area.

The Massif Central, as its name implies, is a mountainous region bridging the gap between the Alps and the Pyrenees, but belonging to neither of them. Much of it is extremely ancient, formed about 400 million years ago, although there are still some grass covered volcanoes the size of Etna and Vesuvius that only became extinct 10,000 years ago. There are cones to be seen in every direction rising in places to 1800m, with crater holes some 40 miles in diameter and flows of pumice reputed to be unequalled anywhere in the world except in New Zealand and Alaska. Generally the region has a base of igneous rock, a crystalline rock of granite created from molten lava, with a covering of sand and clay in the north and limestone to the south, where the Gorge du Tarn is located. The region can simply be described as an outline of a giant cheese riddled with deep cracks, whose edges have been nibbled away by mice for

thousands of years. In the South Central Massif the rainwater and streams drain quickly through the limestone mantle, creating steep sided gorges and a labyrinth of caves and grottos full of stalagmites and stalactites which dwarf the hundreds of visitors who come to see them. The area is characterised by arid plateaux of limestone, lush green valleys, fast flowing rivers, waterfalls and gurgling streams, and villages and towns hewn from the local stone.

I have discovered that for any site there is a particular point, or secret point as I like to call it, from where the new construction can best be envisaged for assessing its scale and relationship with the landscape. When I arrived at the best vantage point I found, not surprisingly, that I stood a short distant from a small church at Brocuejouls, which is half way up the Causse du Larzac some 3km from Millau and 1km from the crossing point. This has happened to me before on other projects. At some point during the history

of Millau and the nearby hamlets our ancestors have found a place which gives a wonderful view up and down the valley, the causse and the town of Millau. Many churches are orientated east–west to capture the best light, as the sun arches over the heavens. The light was also going to be critical for the bridge. I think this connection with the past was particularly rewarding.

Shadow, Light and Scale

After studying preliminary design sketches established by SETRA and following this up with an aesthetic review of the site, the bridge form that we felt worked best among the proposed solutions was a cable stayed structure. We arrived at this solution by comparing the quality and sensitivity of the alternative bridge forms. The beam and column options looked squat and monotonous in

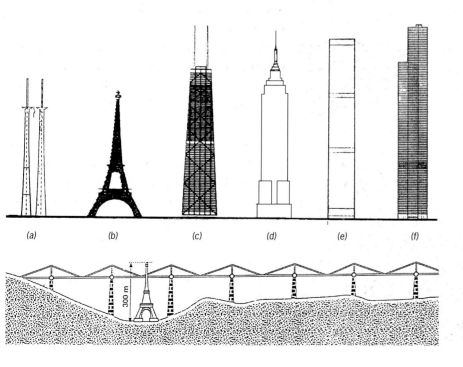

(a)   (b)   (c)   (d)   (e)   (f)

300 m

Left. Comparison of Millau piers with the height of landmarks around the world
(a) Millau piers
(b) Eiffel Tower
(c) John Hancock Building
(d) Empire State Building
(e) World Trade Centre
(f) Sears Tower.

Below left. The overall length of the bridge is twice the length of the Champs Elysées.

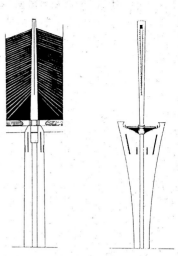

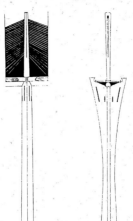

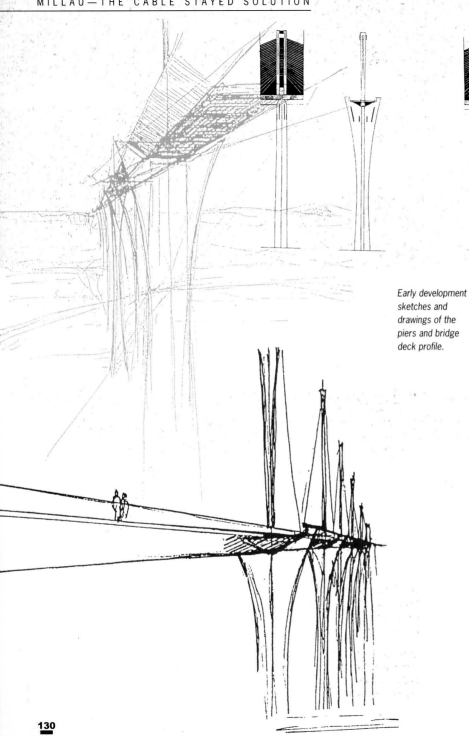

Early development sketches and drawings of the piers and bridge deck profile.

silhouette on the horizon, no matter whether we widened the central span or kept them all equal. Even placing a cable stayed span in the centre tended to flatten the gap of the valley, rather than heighten and accentuate the drama of the crossing.

The cable stayed solution fitted well, giving the best balance of height, scale and tension across the limestone plateaux, the slope of the valley and the spaces in between. We looked at the effect of increasing the width of the main span across the river Tarn from 350m to 750m to reduce the meeting points with the ground. This still looked unbalanced and tended to emphasise the scale of the main span only. Keeping the mast height and pier length the same only increases the squat look of the bridge. It was not graceful.

An equal spacing of pier supports, with a span length of 320m and a fan arrangement of the cable stays, created the most graceful proportions for the bridge. A harp stay arrangement was considered but it did not concentrate the flow lines of the cables quite as powerfully as the fan. One can't help referring to the Ganter Bridge in the Simplon Pass in Switzerland to make this point.

We then carried out a lot of refinements, modifications

and shaping of the concrete pier supports and cable masts to improve the proportion and composition of the bridge elements when seen from different vantage points: for instance from the top of the causse, from the river bank, from Millau, the church at Issis and from the roadway itself. The aesthetic lines of the bridge must be clearly expressed and understood as an integral part of the structure.

We must find harmony between strength and mass, between simplicity and ease of construction and the need for an elegant and sober structure. Paradoxically, when we modelled the piers to look slim, we actually increased the weight of the piers—not quite as dramatically as the piers for the Roquebillier Viaduct at Cahors, where the split of the support almost doubled the weight. An elliptical shaped pier with its main axis at right angles to the bridge line creates a sleek taper when the bridge is seen across the valley. Seen from another angle the pier is a wide sculptured shape that appears to swallow the bridge deck as it widens to meet it. The elliptical cable mast rises above the deck, with the wider section that anchors the cables facing the down carriageway.

It is this visual reception, the play of light and shade on the surfaces of the bridge, that creates the look of lightness,

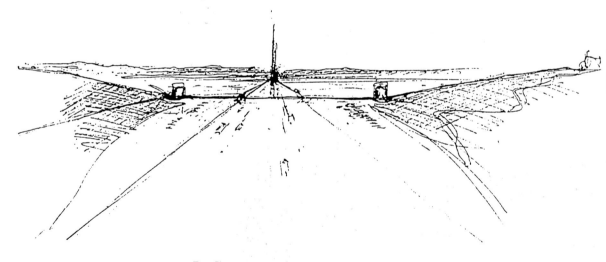

Top. *Photomontage and sight line projections help to simulate scale effects.*

Above. *A sketch perspective of the landscape just before arriving at the bridge.*

not a mathematical calculation arrived at by some empirical science. Often it is a small change of angle or curve or slope that can have a major effect on the bridge composition. Sketches and sight line projections were drawn to help simulate the scale effect of the bridge against various markers on the landscape and horizon.

A Place for People

We wanted to link the motorway to the bridge and the bridge to the local area. To create a gateway before arriving at the crossing point we have designed a large elliptical island of grass to expand the view and to signal the arrival at the crossing. The island gateways mark the entry and exit points of access to the local roads down to Millau, or the bridge viewing area, the river walk, the restaurants or proposed motel village.

In order to limit any negative feelings for the motorist travelling some 240m above the valley for a distance of 2km along the bridge, we introduced bridge parapet screens with perspex panels that allow views of the valley below. The parapets curve towards the deck and are made high enough to increase the security of the motorist looking ahead and along the carriageway. We created emergency stopping points at each pier support with telephones points for emergency services, an access doorway leading to an emergency staircase and lifts for evacuating people or an individual from the bridge. There is a 3m wide hard

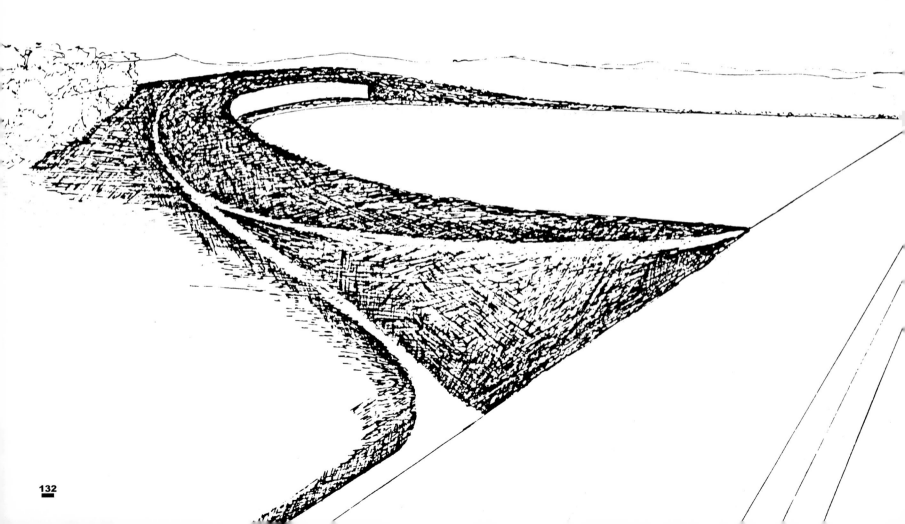

shoulder running the full length of the bridge along each carriageway and crash barriers protecting each side of the 5m wide central reservation, down which the cable mast and cable stays run.

The challenge to design and build an elegant and graceful bridge develops from a spirit of team work and the capacity and commitment to want to build a beautiful structure. Having engendered that spirit in the project, the maxim we use to create an elegant structure is to strive for simplicity and minimalism of form. Attempting this takes a great deal of hard work, careful detailing and repeated refinement of a good design idea…which is rarely simple to do or easy to achieve in practice.

The A14 Nanterre Viaduct Fluidity of Form, Transparency of Design

Kate Purver, Engineer

"I think that perhaps there are three basic modes of perceiving and representing the world around us. They are realism, idealism and expressionism. The realistic modes need no explanation; it is, in the plastic arts, the effort to represent the world exactly as it presents itself to our senses, without attenuation, without omission, without falsity of any kind. That effort is not so simple as it sounds."

Herbert Read, "The Meaning of Art".

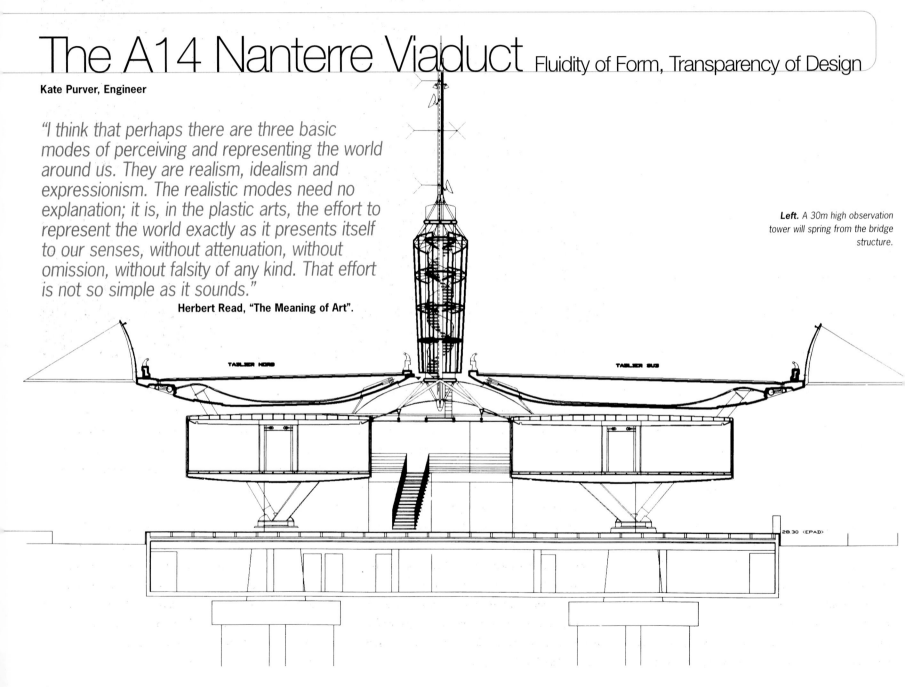

Left. A 30m high observation tower will spring from the bridge structure.

TABLIER NORD   TABLIER SUD

28.30 (EPAD)

Breaking Up Visual Barriers

The Nanterre Viaduct is approximately 250m long and consists of seven approximately equal spans of around 35m. It carries two three-lane carriageways, each on a separate concrete deck whose width varies from 12.5m at one end to 20m at the other. The viaduct is curved in plan. Supported on the two central viaduct piers, just below the carriageway deck, is a building structure and a tower which will rise 30m above the deck.

The viaduct forms a major arterial connection to the west of Paris, linking the bridge at Carrières-sur-Seine to the A14 and A86 junctions near La Défense. The viaduct crosses a stretch of disused and derelict land, which will be transformed into landscaped parkland once the viaduct is complete. The suspended building below the deck will become the administrative centre of the motorway authority. The viaduct and building will be completed in 1996.

The architectural aim of the design was to create a structure whose curved form would appear fluid and slender from all points of perception. The design seeks to break down the visual "barrier" of the bridge bulk by conceiving a structural form that is penetrated by the horizon beyond and by clear air, allowing the parkland below to be seen as a whole.

The deliberate separation of the two carriageways, the use of slender steel arch piers, the curved profile of the deck soffit and the length of the spans all contribute to the transparency of the structure.

The steel piers constitute the most unusual feature of the viaduct. The design evolved as we tried to conserve the transparency of the pier supports as well as emphasise the curvature of the deck soffit and the curving radius of the

Photo: Vincent Hend

viaduct alignment. Each pier consists of a central arch flanked by diagonal struts, supporting both carriageway decks.

The arch, which provides support points for the inner edges of the decks, is a fabricated trapezoidal hollow steel section, 900mm wide and varying from 600mm deep at the base to 1000mm at the crown section. The angled struts at opposite ends of the arch pick up the outer edge of the carriageways, and are rolled hollow steel tubes up to 70mm thick, with an outside diameter of 610mm.

The geometry of each pier is different and has to be asymmetrical in order to accommodate the slope and variation in width of the carriageway decks. The design developed towards the idea of a rigid arch, capable of taking bending under asymmetrical loads, contrasting with slender struts working in axial compression.

The tubular struts are filled with reinforced concrete, to increase their resistance to accidental impact loads. The junctions of the arch and struts are positioned directly beneath the centre of gravity of the carriageway decks. The feet of the piers are connected to a heavy metal base-plate assembly which is cast into a concrete foundation.

Transverse stability of the viaduct (its resistance to sidesway) is provided by portal frame action of the arch. Out-of-balance loads and lateral movement of the carriageway deck are absorbed by bending action of the steel arches. The precise shape of each arch, and of its "shoulder" connections to the concrete deck, was the subject of much detailed analysis in order to minimise arch bending under uniform loading. Excessive bending would distort the arch and significantly reduce its efficiency in resisting lateral movement.

The fabricated "shoulder" pieces are welded to the arches, incorporating anchorages for the transverse prestressing tendons, which pass through the concrete deck, and for reinforcing bars.

The pier foundations are groups of concrete barrettes 15m deep, founded in the Auteuil Sands of the region. Locating the pier base and foundation centreline beneath the deck centre of gravity eliminates the need for a horizontal structural connection between foundations; each foundation can be treated independently.

The Carriageways and The Building

The carriageways run in a continuous gentle curve for the entire 250m length. The decks are of prestressed concrete and are cast as a continuous ribbon over the supports, with no expansion joints. The curved soffit allows the edge of the deck to be as slender as possible and generates a gradual change from shade beneath the bridge to light at its edges. The split carriageway also allows light to penetrate through the central space between the decks.

The deck is 1500mm deep at the mid-section, but its weight has been minimised by the introduction of cylindrical polystyrene void formers. Longitudinal prestressing cables are positioned between the void formers and anchored at construction joints 9m ahead of each pier. The transverse prestress, which takes transverse bending of the deck and the tie force between the arch and the pier strut, is anchored at the arch shoulder pieces.

To minimise intrusion into the clean lines of the carriageway deck, connections for crash barriers, acoustic screens and signs are all cast into the edges of the concrete. Drainage collectors and pipework have been hidden from view, recessed into the deck soffit. Base plates for roadway lighting masts are located neatly on to the crown of the arch

Below. *Plan of bridge layout. The building formed an integral part of the architects' original design concept.*

Bottom. *Cross section through the bridge, showing the building outline. The geometry of each pier is different.*

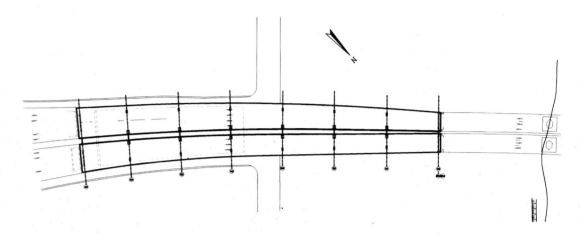

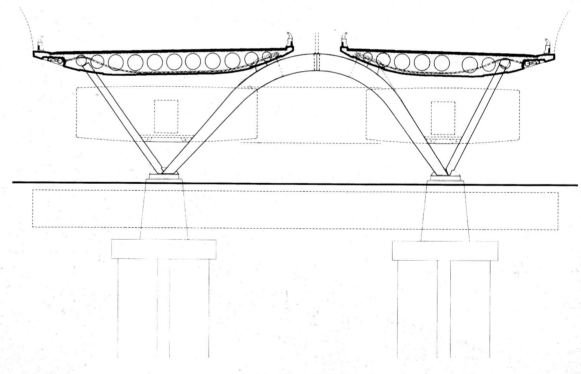

piers, in the space between the carriageways.

Between the central pair of viaduct piers sits the administration building, providing offices for the motorway operating authority. The main spine beams of the building are supported on heavy plate girders fixed between the arch and struts of the piers. Anti-vibration bearings at the building supports prevent transmission of vibrations from the viaduct to the building. There is no connection between the building and the carriageway deck above.

Between the two halves of the building, in the space between the decks, a tower has been designed consisting of a mast and prestressed skin, carrying a spiral staircase leading to an observation platform.

The decision to suspend the building, rather than take the simpler option of designing it as a separate structure, integrates it into the design concept for the viaduct by continuing to allow the passage of light beneath it. Thus the barrier effect is avoided, and the aesthetic quality of the viaduct is maintained.

The innovative design inverts the more usual bridge structural system, of massive concrete piers supporting a steel and concrete deck, replacing it by an emphasis on the lightness of the steel piers and on the fluidity and slenderness of the whole.

Photo: Vincent Hend

A Gateway Bridge for the New A13/A406 Interchange

Ric Russell, Architect and Morris Murray, Engineer

Thoughts rationalising the main issues surrounding the viaduct and the junction have led to certain conclusions. Whilst clearly the A13 is the main arterial route through the interchange, the significance of the junction must be seen as the point of departure providing access to London Docklands, the A406 and Beckton. The notion of a gateway bridge seems highly appropriate in this location.

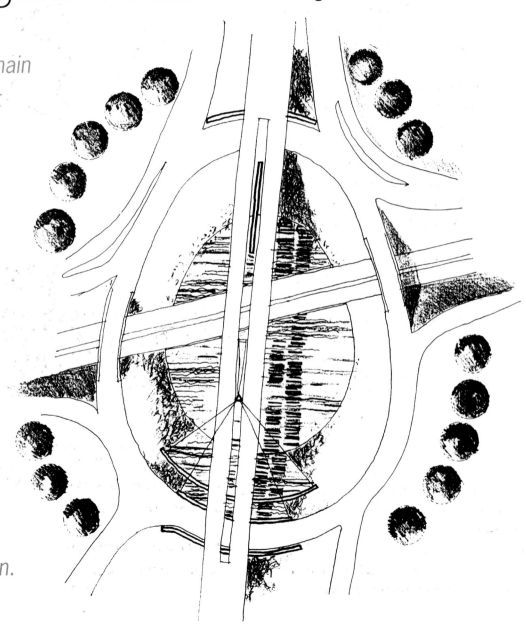

THE ARCHITECTURE OF BRIDGE DESIGN

Notions of Gateways

While the requirements to form a "gateway" structure to London Docklands via Beckton and the A406 were the primary objectives of the brief, the junction also forms a significant reference point on the journey to and from central London on the A13.

Clearly the A13 is and will always remain the dominant and visible roadway through the interchange, with all other routes remaining subservient. In this context the junction at the A406 is a point of departure from the A13 that leads away to various destinations. This suggests that several "gateways" can be served by one centrally placed one. The reverse analogy applies to "gateways" located on the exits. The interchange in many ways should express the positive feeling of arrival and welcome, rather than a negative one of departure, and for these reasons a centrally placed gateway is preferred.

Due to the multi-faceted requirements of the interchange, it is essential that key orientation pointers are developed into the scheme.

The gateway seen from a distance — there should be a clear definition to the driver that allows immediate recognition of the north–south–east–west orientation of the junction and its location. This must work from distances that allow the motorist enough time to negotiate the lanes and slip roads. Of course orientation can be achieved by

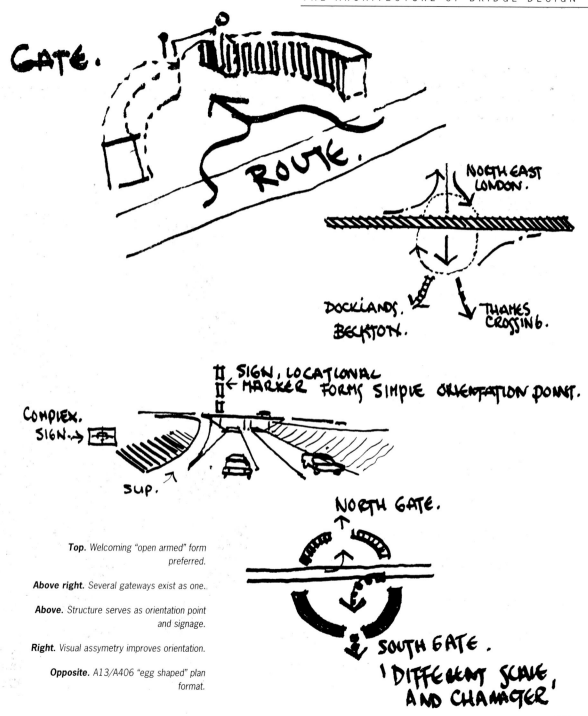

Top. Welcoming "open armed" form preferred.

Above right. Several gateways exist as one.

Above. Structure serves as orientation point and signage.

Right. Visual assymetry improves orientation.

Opposite. A13/A406 "egg shaped" plan format.

139

conventional signage. However, when moving at speed, visual markers of individual character with a "sense of place" are easier to recognise and anchor in the mind.

Asymmetry is better — visually, it would work more successfully if the viaduct structure were not placed central to the A13, or symmetrically about the junction. Asymmetry already exists in the egg shaped format of the roundabout. The centre of gravity of the egg shape offsets the focus of layout to the south of the A13, and this should be exploited in the scheme.

The close up view — the shape of the roundabout creates a strong gateway form to the north and south of the A13. An "open arms" formation is a welcoming one and follows in the tradition of classical gateway structures. This is the gateway form that we prefer.

Sense of Place

To achieve a sense of place that reinforces connections to the Docklands and its marshlands associations, it is proposed that the central spaces of the junction should be flooded. The water pool will add sparkle, light, tranquillity and provide an ever changing setting.

Suggestions for the water feature include a simple pool, fountains, cascading water terraces, underwater lighting, sculpture, bird life and landscaped areas. The creation of the waterpool will mitigate the impact of the overbridges on the roundabout, drawing attention away from them and more towards the enclosure and the pool.

We have concluded that the creation of a simple viaduct crossing over the A13 cuts the area in two and would destroy the sense of place created within the enclosure of the roundabout. Ideally the least intrusive overbridge structure would be one supported by a

EDGE RECREATED AS BRIDGE OVER WATER.

Above. *Introduction of water to provide sense of place.*

Below left. *Multi columned viaduct divides enclosure of roundabout.*

Below centre. *Less structure would be preferable.*

Below right. *Minimal: impact of single support.*

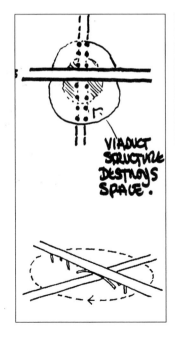

VIADUCT STRUCTURE DESTROYS SPACE.

THE LESS THE STRUCTURE, THE LESS THE INPACT.

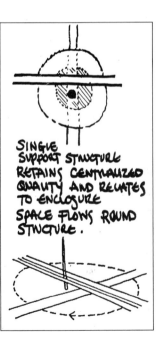

SINGLE SUPPORT STRUCTURE RETAINS CENTRALIZED QUALITY AND RELATES TO ENCLOSURE SPACE FLOWS ROUND STRUCTURE.

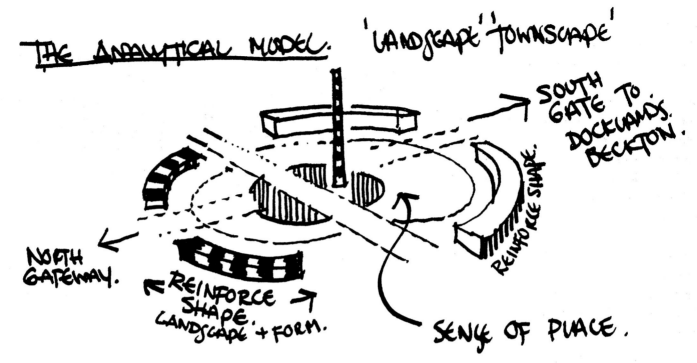

THE ~~ANALYTICAL~~ MODEL. 'LANDSCAPE' 'TOWNSCAPE'

SOUTH GATE TO DOCKLANDS. BECKTON.

REINFORCE SHAPE

NORTH GATEWAY.

REINFORCE SHAPE LANDSCAPE + FORM.

SENSE OF PLACE.

singular stem that rises out of the water pool. If this is not achievable, reducing the regularity of the structure and creating a visual foil to focus attention towards the enclosure of the roundabout should be pursued.

Possible Solutions

Two schemes have emerged, one representing an ideal approach which evokes the sense of place and which defines the need for a vertical marker. The other option is a pure structural approach capable of providing a "central focus" for the flyover crossing the A13.

A single masted, cable stayed solution of the first option responded well to all aspects of context and place. The cable stay offers a column-free span over the A13 and the water pool. To keep the appearance of the structure light, we have split the carriageways and allowed the cable mast to penetrate the space in between.

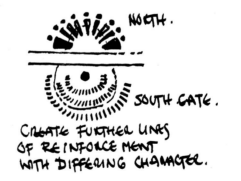

NORTH.

SOUTH GATE.

CREATE FURTHER LINES OF REINFORCEMENT WITH DIFFERING CHARACTER.

LANDSCAPE ENCLOSURE.

Top. *Ideals brought together as model.*

Above left. *North and south differentiated.*

Above right. *Encirclement of roundabout reinforced.*

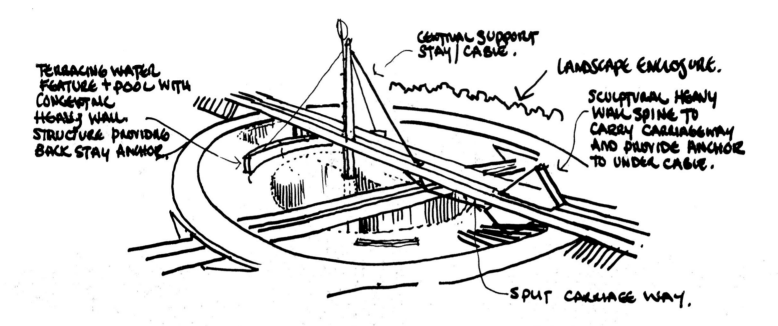

TERRACING WATER FEATURE + POOL WITH CONCEPTUAL HEAVY WALL STRUCTURE PROVIDING BACK STAY ANCHOR.

CENTRAL SUPPORT STAY/CABLE.

LANDSCAPE ENCLOSURE.

SCULPTURAL HEAVY WALL SPINE TO CARRY CARRIAGEWAY AND PROVIDE ANCHOR TO UNDER CABLE.

SPLIT CARRIAGE WAY.

The mast will be stayed by a three cable arrangement and tied back to a heavy wall structure that forms part of the terracing, which is concentric to the water pool on the south. A sculptured wall supports the north end of the span, anchors cables and forms the asymmetrical marker for the northern gateway.

The other solution uses an arched bridge to span over the A13 as the central feature. The bowl shaped roundabout is terraced down to the central water pool and the arch bridge. The terraced walls provide a platform for lighting columns. This option places more emphasis on the A13, creating a sense of place only below the flyover. It lacks the multi directional, all embracing virtues of the single mast solution and therefore would not create the gateway structure we were looking for.

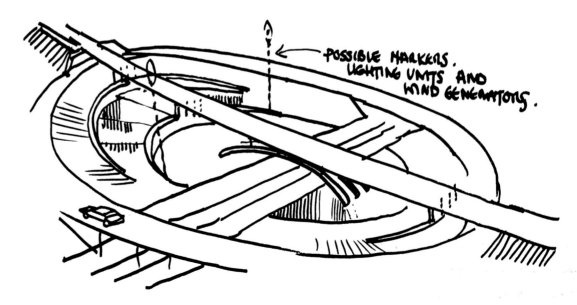

POSSIBLE MARKERS. LIGHTING UNITS AND WIND GENERATORS.

Top left. *Preferred concept with cable stayed structure.*

Left. *Alternative structure considered.*

Above. *Developed concept elevation.*

thefootbridge*explored*

part *six*

Islands in the Water

Ian Ritchie Architects; Ove Arup & Partners, Engineers

A footbridge is first and foremost an environment, a place of experience. Designing a footbridge is an aesthetic activity based on an understanding of context and purpose. This should not be obscured by either engineering gymnastics or technological indulgence. Its design should be informed by the topography over and between which it spans, what it connects to and whom it serves.

Poetry of Place

To walk across a footbridge should be an enjoyable experience whether you are alone, reflecting on the tranquil environment, or hand in hand in a romantic moment. The environment of the bridge should encourage these feelings. The drama of crossing a rope bridge high above a gorge or stepping on stones laid over a stream creates an awareness of the land and the waterscape. The design of the bridge should try to support this experience.

Looking east–west across the dock, the waterscape is dominant. The flat, reflective expanse of water is fringed by blocks of tall buildings. In the future we might expect the calm water to be more disturbed, reacting to the surf from taxi boats and pleasure craft.

On one side of the dock is the hub of the office world with people busily going about their business, or lunching and drinking along the waterfront cafés and bars. On the other side sit motionless fishermen with their carbon fibre rods, tins of bait and the ubiquitous mobile phone. You can see both ends of the dock, but there is no short cut over the water for the people beached on these man-made peninsulas or restricted by the one hour lunch break.

The crossing connects people with places. But this crossing experience is not about the taming of wild waters nor traversing a deep gorge by a structure dedicated to showing off man's engineering ingenuity. It is about creating the poetry of place, of people walking over man-made islands, stepping on stones close to the water.

Below. *There is no point in competing with the Docklands Light Railway bridge, indulging in gymnastic arm-waving.*

Below. *The site is alongside the Docklands Light Railway bridge, the only crossing in a long expanse of dock.*

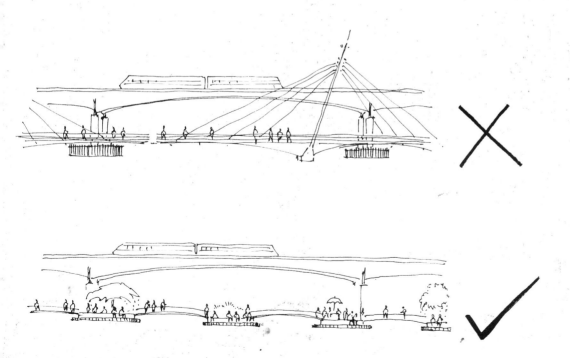

Above. *Although the bridges were initially to be located alongside the rail bridge, the chain of floating islands could be relocated subsequently.*

Visual Context and Location

The blue and red coaches of the Docklands Light Railway rumble on and over the imposing frame of the DLR viaduct, from Canary Wharf to South Quay. The viaduct dominates the immediate environment. There is little point therefore in designing some spectacular footbridge which tries to compete with it. This only increases the visual conflict and adds to the environmental discord of the whole setting.

Our design intent is to create a crossing on a human scale, to create an enviroment that provides pleasure points, places to fish, to walk, to stroll ... and to design it economically.

We wanted to create landscaped islands or eyots across the water, connected by short bridges. These eyots would be designed as playgrounds for human activity, where opportunities exist for fishing, painting, snacking, enjoying the view and the garden themes of the different islands that are crossed. There are sheltered places on the eyots to escape the wind and rain during the crossing.

From east to west the waterscape of the dock dominates, and the string of eyots with trees, shrubs and linking bridges can be clearly seen. In the north–south perspective, the crossing is foreshortened by the stacked rows of eyots and the converging line of the walkways engulfed by the imposing tower of One Canada Square at the far end.

The brief asked for the bridge to be located adjacent to the DLR viaduct. The longer section was to be placed to the east and the shorter length to the west. Neither location is "quiet" visually or acoustically. We suggest that the position of the eyots on the east side — eight of them in all — should be located in a position some 100m east of the DLR viaduct.

We do not want a bridge permanently in the shadow of the overbearing DLR viaduct and within earshot of the noisy rumblings of the train.

Structure of the Eyots

The eyots are constructed by forming elliptical concrete pontoons set 15m apart, that rise and fall with the changing water level in the dock. Each pontoon is tethered to a vertical pile driven into the bed of the dock. The pontoon carries the load from the landscaping, the trees and planters, and has a draft of 5.9m to provide the required buoyancy to float the eyots at a preset level above the waterline. A freeboard of 1m caters for the variation between the maximum and normal loading from foot traffic and bystanders.

To increase the stability of the pontoon, the hull contains ballast tanks of bilge water. The quantity and position of the ballast is adjusted by automatically operated bilge pumps. This ensures that the eyot will float at the correct level.

Because of the pontoon's elliptical shape, there is a

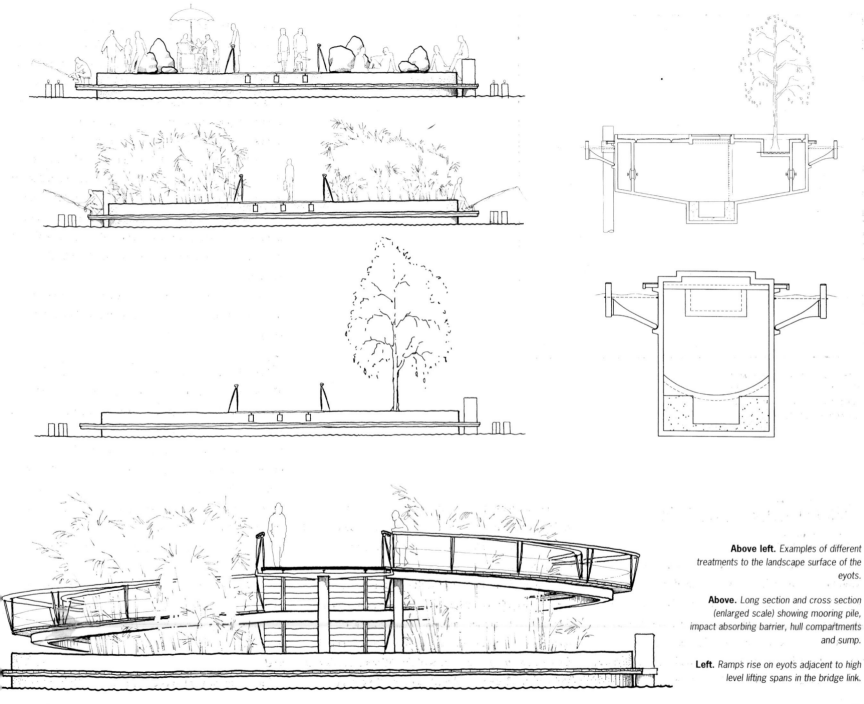

Above left. Examples of different treatments to the landscape surface of the eyots.

Above. Long section and cross section (enlarged scale) showing mooring pile, impact absorbing barrier, hull compartments and sump.

Left. Ramps rise on eyots adjacent to high level lifting spans in the bridge link.

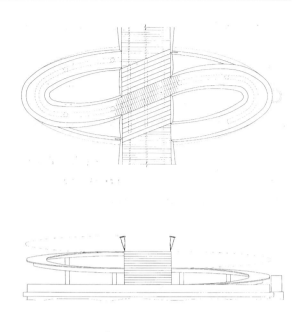

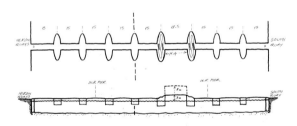

Top. *A ramp and steps link to high level spans.*

Bottom. *Layout of the longer bridge link: the high level span lifts to allow larger craft to pass.*

tendency for it to become slightly unstable across the short width, when high winds act on the trees and sides of the pontoon. The instability is arrested with the help of the rigid construction of adjoining footbridges and the lateral restraint from the connection of the bridge beams to the pontoon.

The pontoon hulls are built of reinforced concrete, with walls typically 200mm thick. Rainwater and seepage from planter boxes and surface soil drain into the bilges. The bilges are compartmentalised by bulkhead walls. Automatic sump pumps maintain the correct status of the pontoon ballast during inflows or significant changes in pedestrian loading. The bow and stern of the hull are also isolated by bulkhead walls. This ensures that the pontoon will remain afloat should the bow or the stern of the hull be breached accidently.

The visual character of the eyot—what you see above the waterline—can be modelled as a repeating design or changed to express different contextual themes. A combination of dwarf bamboo greenery gives a coherent and appealing image to the eyot landscape. This repeating pattern may be broken up by eyots supporting a tree, a timber terrace, a grass bank, a sculpture of rock formations or just a flat timber boardwalk. The opportunity exists for collaboration with imaginative landscape artists in the design of the eyots.

Low maintenance is considered fundamental to the choice of the landscape treatment.

Walkway and Lifting Bridge

The walkway structure spans 12m between pontoons and consists of transverse timber planks supported on three 200mm by 300mm deep rectangular hollow beams. The timber character of the deck is emphasised by leaving the cantilever ends of the planks free of clutter. The parapet takes the form of a shallow lattice girder, comprising a series of wishbone struts which are tied at the joints by interconnecting steel tubes. The parapet is attached to the bridge deck at four points. At the ends of the spans the wishbone struts are inverted to form an A-frame making a rigid end restraint for the parapet.

Only the end spans of the nine footbridges are hinged where they connect to the dock wall. This allows the bridge and eyots to move independently of the dock wall. Otherwise all bridge spans are rigidly connected to the eyots.

The pontoons which support the navigation span house the lifting mechanism to allow boats to pass beneath them. Hydraulic rams in the pontoons raise the interconnecting walkway above the steps and elliptical approach ramps that lead to the walkway from the eyot.

The lifting spans are articulated as three pin arches. The hydraulic rams are surrounded by telescopic guide rails which protect the rams from lateral forces and stabilise them if any sway or eccentric movement occurs. The bridge walkway rises very slowly, balanced by a counterweighted gate which lifts a metre above the landing edge. The walkway continues to rise to 8m above the waterline.

An Ecological Design

The concept for the bridge allows it to be moved, repositioned, and replaced in an easy and efficient way. The island pontoons are made of durable concrete which are easily formed. The walkways are made of North American oak from a sustainable forest source.

Each pontoon has a wide timber fender allowing people to sit or place their feet on them. Beyond the fender and located just below the waterline is a safety net of woven polymer.

The walkways all run in a straight line with a very gentle arch. The pontoons are elliptical in plan. At night, fibre optic cables light up the walkways and the pathways across the eyots. The safety net just below the waterline is lit via a submerged fibre optic cable which runs along the outer face of the fender. This creates a halo of light on the perimeter of each eyot. Areas of landscape and seating are also floodlit.

This bridge is easy to dismantle and to re-site. The pontoons can be moved individually or in a string, just like barges, and towed across any calm stretch of water. It is a wholly transportable bridge with a high re-use and salvage value.

Investing in a new bridge is expensive. Shouldn't they all be made recyclable, like this one, with no inherent redundancy and little structural maintenance?

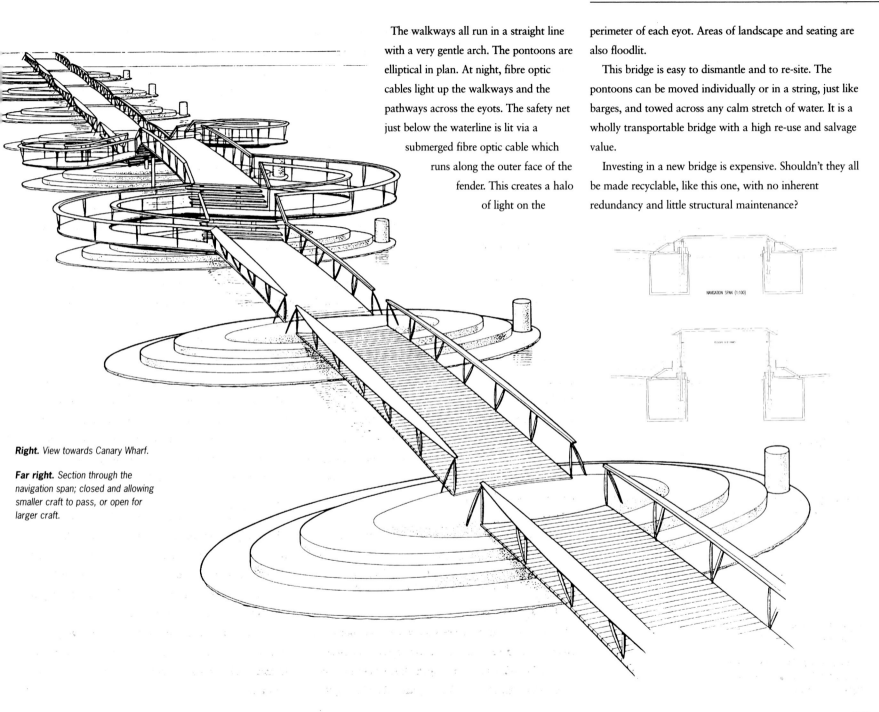

Right. View towards Canary Wharf.

Far right. Section through the navigation span; closed and allowing smaller craft to pass, or open for larger craft.

NAVIGATION SPAN (1:100)

La Villette's East Footbridge

Alexandre Cot, Engineer

The corkscrew twist of the structure would not look out of place on a roller-coaster ride, except that here we are trying to keep the walkway as level and as stable as possible. The structure of the footbridge was designed to express the wildness of La Villette park into which it leads.

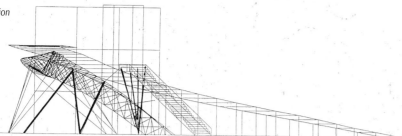

Right. The design features a juxtaposition of curves and sharp angles.

La Villette Park is an architectural garden designed by Bernard Tschumi around the focal point of the highly acclaimed Cité des Sciences technology museum. The east footbridge is one of two crossing the Canal de L'Ourcq which runs through the park.

The east footbridge forms a 45m long walkway connecting the south side of the canal to the museum, which is situated in the north part of the park. The line of the bridge was partly dictated by the need to maintain clearance of 6m above the canal for navigation.

The architect wanted the footbridge to express the wildness of the park he had designed. As the engineers, we have designed a structure which attempts to convey this impression. Our approach was to design the bridging structure with a degree of independence from the walkway it supports. The continuity of the walkway, with its restrained and flat profile, works in contrast to the twisting form of the supporting structure.

The structure was designed as a triangular truss of welded tubes, which could curve in plan and in elevation along its entire 45m span. The geometry of the truss rotates gradually about its centreline, which follows an ellipse in space. While the truss enjoys this geometrical freedom crossing the canal, the walkway retains a sense of stability and calm.

To emphasise this difference in attitude between the walkway and the truss, a deliberate gap was maintained between the two elements. The walkway is connected to the support truss via short steel struts to keep the gap as visible as possible. The deck is built of relatively heavy precast concrete planks 160mm deep, supported on concrete ribs every 3m. The contrast of a lightweight truss holding up a heavy walkway adds to the sculptural interest.

The supports on each side of the bridge were subject to a number of construction constraints. The canal embankment on the south side did not allow enough space for wide foundations, leaving room only for point supports on single piles. On the north side, although there was adequate room, space had to be made for fire engine access. Consequently

Right. The bridge crosses the canal some 6m above the water.

the bridge sits on simple supports at the south end and a portal frame at the other. The look-alike supports at both ends were designed as inclined struts of the same diameter, but with slightly different inclinations.

Under self-weight there is therefore an out-of-balance horizontal force acting at the top of the supports, along the axis of the bridge. This thrust is resisted by a pair of tie rods on the north side. Because the stability of the bridge is thus ensured by just two rods in tension, it is critical that this force is never reversed (the slender rods would be useless in compression). So the walkway structure has been made extra heavy using concrete, to maintain the rods in tension under any load conditions.

At the southern end of the walkway a staircase leads down to the canal. The approach from the north end is via a 60m inclined ramp which follows the same curvature as the footbridge in plan. The concrete deck of the walkway is repeated on the ramp, which is supported on slender steel struts every 3m so as to be perceived as a ribbon supported on matchsticks.

The Leith Footbridge

Peter Clash, Architect and Matthew Wells, Engineer

This design for the new footbridge at Leith aims for a strong statement and an identity which will make a new context for the shore area of Leith. Leith was once Scotland's principal port and a centre of great wealth when timber shipbuilding was in vogue. The current revitalisation of former industrial sectors for residential, leisure and retail development has expanded to the dock side and shore areas. The new bridge provides a pedestrian access leading down to the quayside to complement the walkways that have been planned around the docks.

The footbridge curves on plan and rises in a shallow arc making oblique connections with each bank. The stronger convex curve of the inside edge hides the connection with the opposite bank, as you walk the span. The principal architectural objectives are directed towards the manipulation and control of the sightlines and outlook of the pedestrian crossing the water.

Left. *Section along the line of piles showing "leaf spring" column heads.*

Below. *The parapet rises above the eyeline.*

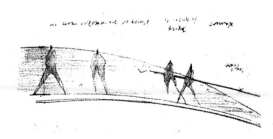

Leith has suffered from the stop–go politics of planning control; the tell-tale signs of overzealous development in the 1960s has left its mark. The context for the bridge is across a body of water overlooked by buildings ranging from industrial sheds to commercial buildings unconnected with any dockland industry or activity.

The use of timber for the bridge structure is a deliberate reminder and reference to the historical associations of shipbuilding in the area. However, the formal resolution of the design has not been taken as an excuse for nostalgia, but rather as a belief in the future of Leith itself and an expression of modern bridging ideas.

Constructionally, the bridge is a complex three dimensional form, developed from the requirements of both structural necessity and architectural intent. The proven technique of glulam and laminated timber has been extended to create three dimensional shapes and sections. Gluing small sections of timber together, rather like plywood, improves the strength and dimensional stability of the finished timber. It is less prone to warp and very tough. The mouldability of timber allows the profiling of laminated sections to be finished as a solid face on one side and finely detailed joinery on the other, with seating and balustrading.

The exceptional strength to weight ratio of timber has been exploited to develop a lightweight structure with a long span, which is sufficiently robust to sustain all the anticipated loads. The ends of the main bridge beams are flexible and are set slightly above the end support walls. By interlaying vulcanised rubber in the glulam beam over the end section the structure responds with the feel of a spring, deflecting under the weight of a pedestrian, then returning back to its original level.

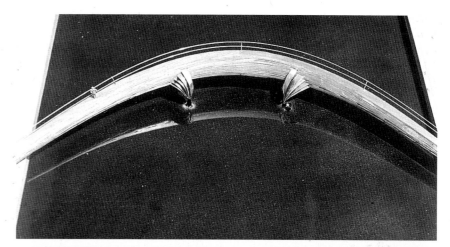

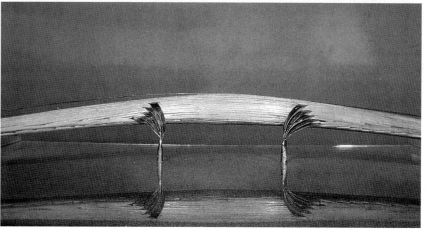

Top. *Plan view.*

Bottom. *View from west.*

The centre of gravity of the boomerang shaped bridge allows the pier supports to be positioned almost on the external edge of the inner curve. The piers' eccentric position helps to restrain the tendency for the span to twist and also improves the torsional stiffness. The piers are formed from tubular steel driven piles with column capital of steel plates, like leaf springs, connecting the pier to the bridge deck. The upper end of the plates are splayed to offer a large contact area with timber beam. Two timber splices are made in the main beam to connect with the steel plate ends.

Each pier is protected from accidental damage by a floating boom. Lateral restraint to the bridge is provided by the tubular steel piles and the connection arrangement of the bridge beam to the banks at each end.

The footbridge curves on plan and rises in a shallow arc, making oblique connections with each bank. This is an intrinsic part of its architectural statement. In elevation the beam depth increases to a maximum at the centre of span, while the walkway deck also widens to its maximum at mid span. The shape and curvature distort perspective, to emphasise the curve of the bridge and the structure of the beam. The two piers emerging like hands out of the water, divide the 50m bridge into three spans. The strong convex curve of the bridge keeps the pedestrian eye trained away from the exit point on the opposite side, and encourages an outlook towards the open body of water and eccentricity of the crossing.

The detail of both parapets sets out to manipulate and deceive the line of sight. The rising arc of the solid parapet wall on one side and the increasing width and splay of the balustrade rails on the other, make the bridge appear shorter when seen from the bank. Seen from the bridge, the span looks a lot bigger. We wanted the pedestrian to find an equilibrium on reaching the middle of the bridge, a central place with wide balustrades on one side for leaning on and seating on the other to watch the day go by.

A design like this can be shaken out of a body of conventional, code abiding design manuals. A choice can be made using criteria which include economy of

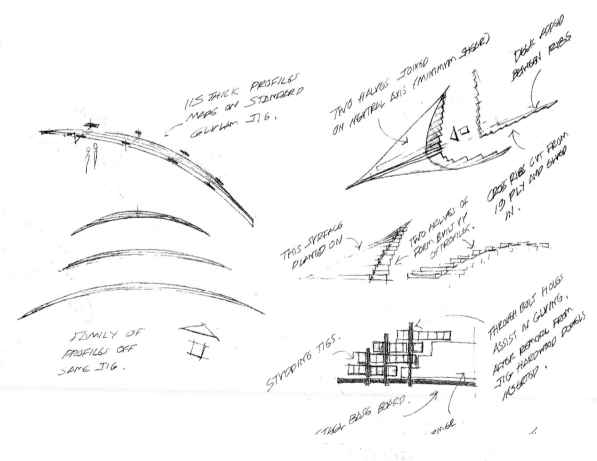

materials, labour intensity, finance cost, maintenance and life cycle costs and construction method. A structural concept like this should not be an unusual event, as it does account for all these economic building factors. What is obviously low on the list of priorities when commissioning bridge designs, is the lack of commitment to appoint designers with a singular and individual vision.

Below. *Construction sketches.*

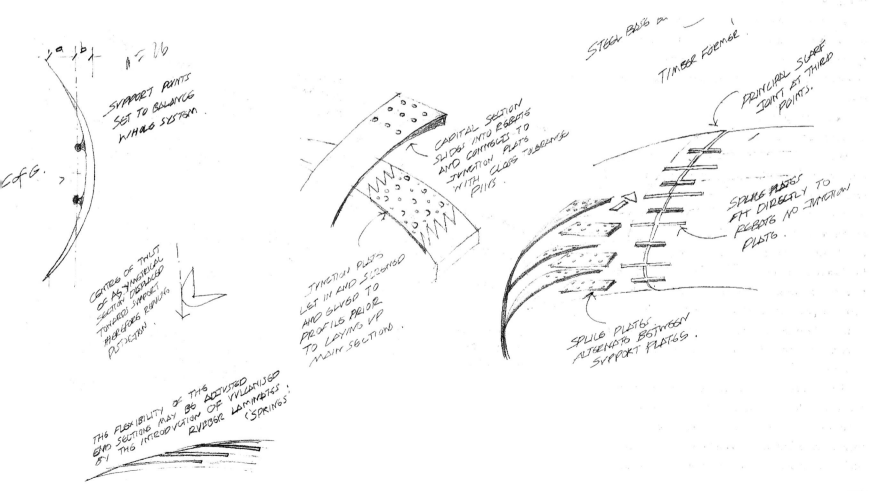

The Albert & Victoria Footbridges the Bridge as Art

Cezary Bednarski, Architect

The Albert Dock In Liverpool and the Royal Victoria Docks in London are both spectacular and unique urban spaces dominated by water and in much need of a new soul, to link them with the cultural and social activities of the area. A bridge can serve as the link, a place for people to enjoy, a place endowed with civic qualities, that can be sculpture, art and a utility all in one. A bridge can put dramatic expression and breathe life into a sterile landscape.

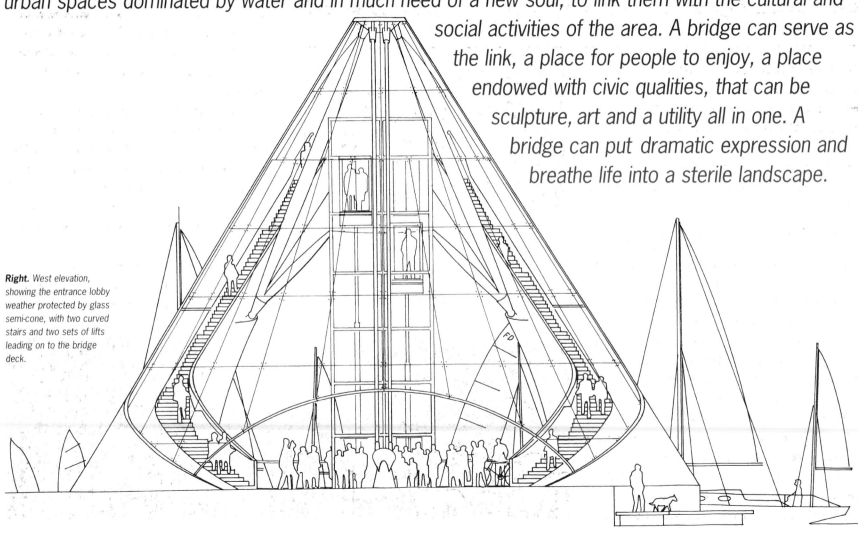

Right. *West elevation, showing the entrance lobby weather protected by glass semi-cone, with two curved stairs and two sets of lifts leading on to the bridge deck.*

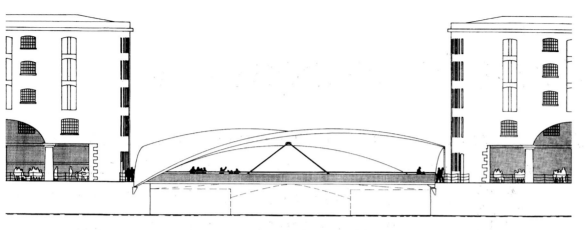

which holds up the walkway. It is fabricated out of two seamless surfaces made of aluminium sheeting, held apart by aluminium channels of varying depths. The canopy shape will be mathematically defined as two intersecting hyperbolic paraboloids or a toroid. Its shape alludes to the gentle roll of a wave, of the surf breaking on the shore, a section of a giant shell, an unfurled mainsail, ... and in the spirit of the docks' past building tradition.

The canopy construction is a continuation of the creative engineering tradition of the docks where the roofs of the dock buildings are a series of riveted iron plates which act structurally with the slender iron roof trusses, to hold themselves up.

The canopy will be supported on four pin joints set off the dock wall, so that no fixings are made to the adjacent buildings nor within the narrow sidewalks of the dock. The lightness of the entire construction and absence of lateral thrusts permits very slender supports. The walkway deck of toughened glass 'floats' within a perforated steel trough which is suspended from the canopy and supported on to the dock walls.

The bridge is a poetic insertion into the strongly articulated masonry and iron construction of the pavilions: an insertion and a landmark visible from the city, a piece of

The Victoria Dock bridge was designed before the Albert Dock footbridge, but in explaining the idea behind our design, it makes sense to describe first the Albert Dock footbridge. While the Victoria has a span of 150m, the Albert has one of 44m and they are located poles apart; nevertheless, there are parallels in their conceptual design approach: parallels which suggest some universal imperative on footbridge design.

The Toroid Sculpture

Once variously described as powerfully poetic, solidly stoic and nobly forlorn, the Albert Dock in Liverpool no longer vibrates with the presence of stevedores, the rattle of winches and crane hooks, the aroma of cargo from faraway places, the comings and goings of trolleys and freight. The recently refurbished dock buildings stand stark, devoid of a public statement despite the presence of the Tate Gallery, alas, hidden behind its blue panelled facia.

The pedestrian bridge we were asked to design, was to offer a covered walkway to the visitor over a 40 m stretch of water, between the Edward and Atlantic Pavilions. The proposed design is an attempt to bring back to the docks the excitement of the unusual and the unexpected.

The bridge comprises a load bearing sculptural canopy

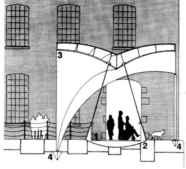

Above. Elevation as seen from Albert Dock.

Left. Cross section at centre span/ suspension point.

Below. Computer simulation as seen from Albert dock.

art and sculpture, belonging unequivocally to the present time. The bridge will have its own light and acoustic environment.

Spatial Art

At the outset we felt that a footbridge spanning over the Royal Victoria Docks in London's Docklands, which has a busy yacht club and which stages high speed boat races, should capture some of this character.

In essence the bridge structure comprises two stiffened cantilevers—which are anchored by a system of struts and ties to caisson foundations at each end.

We have designed the footbridge as a place for people to visit rather than just to pass through; i.e. as a place endowed with civic qualities, similar to those of a gallery or exhibition hall.

The footbridge, elevated 15m above the waterline and spanning 150m, has been conceived as a series of spatial experiences. There is an arrival area on the dockside, a glass

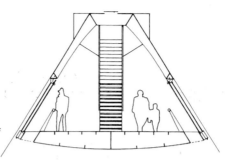
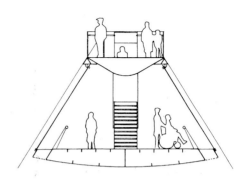

Right. Cross sections through bridge deck.

Below. Computer simulation showing proposed bridge as seen from west end of Victoria Dock.

walled semi-conical entry enclosure with two curved staircases and glass walled lifts, the elevated deck itself and fuselage enclosure which opens on to the fully glazed section of the centre span and a high level viewing gallery or poop deck.

The cantilever construction of the bridge was chosen for both structural and aesthetic considerations. Plate technology has been the tradition of ship building in the area and for the gantry cranes which now stand motionless, just a decoration on the dockside. The cantilever sections that support the

walkway are built up from stiffened steel plates formed into inverted U beams, in a tapering profile. The deeper profile of the U beam at the support end carries the gently curved staircases that lead to the walkway. The cantilever is pin jointed to the concrete foundation base. Circular hollow struts and steel rod ties, anchored to the foundation base, connect the upper flanges of the cantilever beam to the base.

We felt the bridge should have strong horizontal planes to contrast with the tall shape of the cranes and the lattice frame

By night the lighting of the bridge has bee

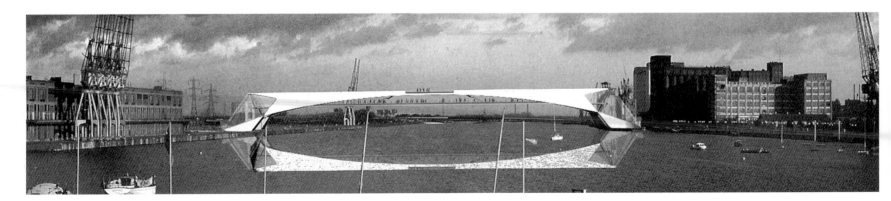

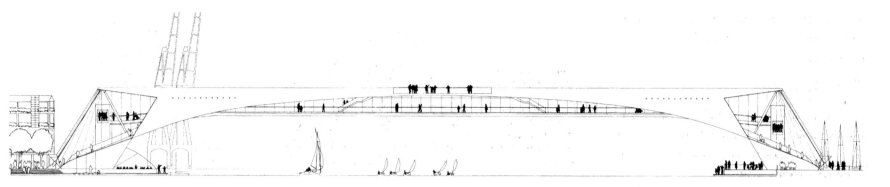

of the exhibition centre which it leads on to. The sculptural form of the cantilever is picked out in a bold paint colour to emphasise the horizontal line and structural principle of the bridge.

The lighting of the bridge has been designed to express the structure as sculpture also by night. It reinforces the powerful horizontal lines of the bridge, turning it into a celebratory landmark and a counterpoint to the illuminated pyramid of the Canary Wharf Tower. The detailed design was to explore the possibility of colour washes on the cantilever section and the underside, to create a connecting ribbon of light with a clear reflection in the water.

If we agree with Isamu Noguchi that "sculpture is a vital function of our environment" then we must agree that sculpture should be an intrinsic quality of bridges. We follow this philosophy in all the bridges that we design.

Above. *West elevation.*

Left. *Computer simulation of opening celebrations.*

Below. *Plan as seen from above.*

esigned to express the structure as sculpture

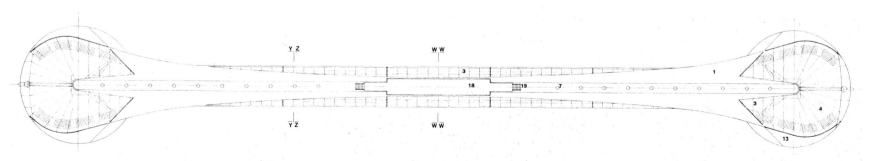

The Cornmarket Canopy Bridge

Ian Ritchie Architects (Simon Conolly Project Architect), Ove Arup & Partners, Engineers

The structure and covering of the canopy and bridge provide an elegant, lightweight solution which affords protection from the elements without significantly reducing daylight. A footbridge across the river is a fundamental part of the scheme. The canopy and bridge will forge a physical and tourist link between the commercial centre of Patrick Street and the historic Shandon area of Cork.

The Cork Historic Centre Action Plan identifies Cornmarket Street and Castle Street as hinge points at the heart of the study area and the point from which the Historic Centre can be integrated into the main commercial sector of the city. The plan goes on to say that the key aim of the development is to provide a viable open air market, which can function seven days a week.

To achieve this aim we consider that a dramatic unifying scheme, which will completely change the vista of the street, is the most likely way to galvanise interest from all concerned parties. The scheme has been prepared to help stimulate a symbiosis between the public and private sectors in the area of planning implementation and funding.

The project explores the physical realisation and fulfilment of several objectives identified in the action plan. A glazed canopy constructed centrally on Cornmarket Street, with open air trading beneath, provides pedestrians, tourists, shoppers and traders with shelter without reducing daylight . Limited traffic access will be provided on each side of the canopy to service the trading staff and existing properties in the street. Pedestrians will be favoured over cars by omitting kerbs and providing bollard protection between walkways and roadway areas. The walkways under the canopy of the trading stalls and footbridge will be paved with smooth flagstones while the vehicle areas will be paved in limestone setts.

The animation of the river, with activities on the bridge, can encourage recognition of the river as a focus for water borne interest. The structure of the footbridge reinforces the link across the river to Shandon and areas to the north. The steel skeletal form evolved for the canopy structure of the market stalls is echoed in the bridge truss.

The structure designed for the footbridge and trading stalls, comprising a painted galvanised steel frame with cast nodes and toughened glass canopy, is capable of a number of variant forms that can be used elsewhere in the City, as required.

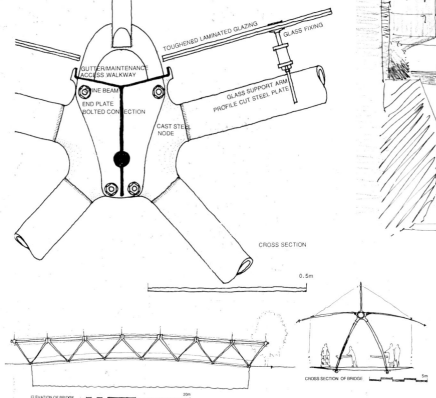

CROSS SECTION

0.5m

ELEVATION OF BRIDGE

20m

CROSS SECTION OF BRIDGE

5m

Above. *Plan and view from Shandon Hill.*

Left. *Market canopy structural form is related to the bridge truss geometry; the arms are lowered to shed rainwater into the river and provide more shelter.*

Japan Bridge

Kate Purver, Engineer

The Boulevard des Bouvets cuts through the overspill district of Valmy, adjacent to La Défense, separating two blocks of offices by seven lanes of traffic. The bowstring arch and glazed canopy of Japan Bridge sailing over the seven lane highway brings a visual sense of order and calm to the chaos below, offering the pedestrian safe passage across the road.

Fine Tuning

Japan Bridge was initially conceived by Kisho Kurokawa, the architect of one of two office buildings which face each other, 100m apart, across the main east–west road out of La Défense. Kurokawa envisaged a low arching tube structure, in the shape of a traditional Japanese bridge, carrying a pedestrian walkway spanning between the two towers. However, it was soon clear that the supporting buildings could not take the horizontal thrust of an arch. This led to the development of a tied arch or "bowstring" solution, which modifies the structure but retains

the original arch shape sketched by Kurokawa.

We also introduced the idea of the double arch: two arches leaning inwards and converging at their summit, stabilising each other. By providing a wide base, this also resolves the torsional instability problem which would occur with a single arch. The geometry was complicated by the asymmetry of the supports; the two buildings are not parallel. The architectural emphasis on visual coherence led to an asymmetrical design,

with one arch longer than the other, and element lengths and angles changing gradually along the bridge. Although the two arches appear identical, the result is in fact that no two elements of the bridge structure have the same length nor the same connecting angle. Computer techniques were essential for generating the geometry, for analysing the structure and for fabrication.

The tendon and deck structure is suspended from the arches by hanger rods at approximately 4m intervals. These are prestressed by the weight of the structure and thus create a stiff, stable plane between arch

Above. The separation of the walkway from the tendon contributes to the visual lightness.

Left. Detailed analysis was undertaken to investigate the buckling behaviour of the arches.

Far left. The architect's original sketches showed a low arching tube.

How do we position a 100m span bridge, 15m above a busy seven-lane highway without closing all the roadways, without recourse to lots of storage space, cranes, a giant helicopter?

and tendon.

The deck structure is a triangulated truss made up of bolted H sections, supporting precast concrete slabs. A glass strip along either side of the concrete gives the pedestrian a view of the activity below the bridge, drawing attention to both the lightness of the bridge and the relative calm within it. The walkway is enclosed by a curved glass canopy held in place by arched stainless steel ribs. The canopy has air gaps for natural ventilation, to reduce the "greenhouse" effect on sunny days.

Considerable study was devoted to defining the geometry of the arches, since calculations showed their structural behaviour to be very sensitive to their precise shape. The arches follow modified parabolic curves, chosen to correspond as closely as possible to the distribution of weight along the bridge. This minimises arch bending and allows the arches to be as slender as possible.

Stability under lateral wind loading was critical. As the wind load increases, the forces in the windward arch and tendon decrease until at a certain load the hangers would go slack. The stabilising effect of the hangers would then be lost, and the arch would buckle. In order to prevent this happening, modifications were made to the structure, including changing the

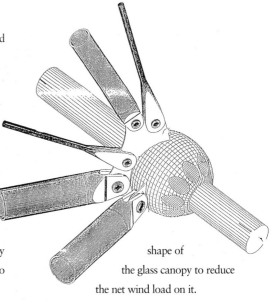

shape of the glass canopy to reduce the net wind load on it.

Construction Finesse

How do we put up a 100m span bridge, 15m above a busy seven-lane highway offering no space for storage or cranes, without closing all the roads or resorting to a giant helicopter? Throughout detail design, consideration had to be given to the difficulties of construction on such a congested site.

The arch itself is a triangular steel hollow section 900mm deep, welded from plate between 30mm and 55mm thick. The tendons are 200mm solid steel rods. The principal nodes — which connect the tendon, hangers, horizontal spacer struts, deck support struts and bracing — are of solid forged steel. The nodes were forged in two halves, each half threaded on to 4m lengths of tendon before being bolted together on site. Each is a standard hemispherical element, with connection plates welded on at the varying angles required by the geometry. Particular precautions were taken in detailing the tendon-to-node connection to reduce the risk of brittle fracture, a phenomenon which can cause failure in tension at low temperatures.

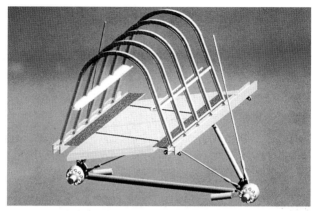

Opposite far left. *The bridge retains its graceful curves when seen from below.*

Opposite left. *The main nodes connect the hangers, deck support structure and bracing to the tendon.*

Left. *The contractor generated three dimensional computer models to illustrate the final appearance of the walkway structure.*

Below. *The bridge spans 100m over densely trafficked roads. At night, lighting illuminates the glazed walkway from within.*

The contractor's solution to the problem of construction was to erect a temporary scaffold bridge, with supports squeezed between trafficked parts of the roadway. The arches were fabricated in up to 40m lengths, brought to site on special articulated lorries and trailers and erected during overnight lane closures. Each section was lifted into place using two mobile cranes and supported on temporary scaffold towers. These lifts were a particular challenge for the contractor to overcome, as precise positioning of the cranes was vital.

Once in place the arch sections were welded together and the tendon and deck structures then put in piece by piece. The whole of these structures had previously been test-erected in the contractor's workshop, in order to ensure that every connection plate was welded at the correct angle and that all the elements fitted together within the required tolerances.

Finally the temporary arch supports were jacked up, to a height predetermined by analysis, to enable the final connection to be made.

Completed in 1993, the project was begun by Peter Rice and finished by his Paris practice RFR. The structure illustrates how architectural and structural principles can work in harmony, with the red curve of the arches providing an expression of the bridge's geometry and emphasising its purity of line.

Canary Wharf Footbridge — Viking Raider

David Kirkland, Architect and Richard Craig, Engineer

The bridge design was a submission for a competition sponsored by London Docklands Development Corporation. The bridge was required for a series of three pedestrian crossings over a stretch of water separating Canary Wharf and Heron Quays from South Quay Plaza and West India Dock. Half the bridge had to be temporary until Heron Quays was fully developed, while the other half had to be designed so that it could be relocated to a new position with relative ease and economy.

Far left. LDDC cranes.

Centre. (NGP) Waterloo retail scheme.

Left. Nuam Gabo—"Constructed Head No. 2".

Analogies and Inspiration

The bridge is located in the heart of London's second financial district, the Canary Wharf Development, and sits in the shadow of the gleaming tower of One Canada Square. It will also be dominated by the massive form of the Docklands Light Railway viaduct and the lesser tower blocks and buildings that surround West India Docks and South Quay.

Our design concept for both the permanent and temporary sections of the three linking footbridges was inspired by the nautical heritage of the area and a desire for an elegant yet affordable structure. The basic requirement of getting pedestrians and cyclists across fairly short reaches of water meant that we could keep the structural form of the bridge simple.

For the shape of the bridge spans and choice of construction material, reference was made to the structure of a boat hull. For the temporary spans we used entirely timber construction.

We were very aware that a footbridge is examined at close quarters. Keeping the construction of the span as simple as possible allowed us to put a big effort into increasing the quality of the finishes of those parts of the bridge that are touched and seen at close quarters—the balustrading, the walkway flooring, the reflections of the bridge on the water by day and by night.

For this reason the permanent bridge spans were designed using steel plate, which could be polished to bring a gleam and sparkle to the reflected quality of the spans. Both the permanent and temporary spans have the same hull profile, contrasted only by their material properties and expected design life. The steel span had to last 50 years and the timber about 15 years.

The steel span comprises a series of interconnecting fins, acting as longitudinal beams, which contour the outline of a boat hull. We drew our inspiration from the sculpture by Nuam Gabo called 'Constructed Head No. 2', where the impression of a solid form of a head and torso was created by plates of metal.

The requirement for the bridge to open has been exploited to enhance the overall form of the hull-like spans. The sweeping tail of the counterweight ensures that the lifting section of the bridge is clearly visible and adds dynamism to the crossing.

To highlight the hull profile of the bridge by night, the spans will be floodlit from the piers. This will also accentuate and contrast the rounded form of the timber spans with the hard edged, gleaming lines of the steel. The walkway lighting is a continuous source running along the handrail.

Engineering Approach

Both the temporary and permanent footbridges are a series of 20m spans supported on reinforced concrete piers. Each pier is founded on a pair of 1m diameter driven tubes, 20m long, which are bedded into the underlying Thanet Sands.

The steel spans consist of a metal deck plate acting compositely with a series of five longitudinal steel plates or fins, which vary from 1m deep at mid span to 200mm at the supports. The fins and deck plate form T beams, with the fins acting mostly in tension and the deck plate in compression. The 20m long fins are stabilised against buckling by steel diaphragms running transversely every 4m.

The timber spans have been designed to construction principles used in boat building. Two primary spine beams, made from glued laminated timbers, are connected to diaphragm ribs to form the carcass of the hull. An outer skin of plywood, supported and strengthened by timber stringers, covers the hull. The shaped diaphragm ribs and timber stringers maintain the profile of the hull. The strength of the bridge comes from the primary spine beams and the plywood outer skin, acting compositely.

Both temporary and permanent bridges require a lifting span for navigation purposes. We designed a bascule span

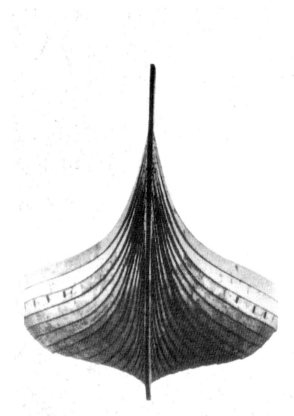

Above. Canary Wharf Footbridge.

Above left. Longboat.

Below. Heron Quays elevation.

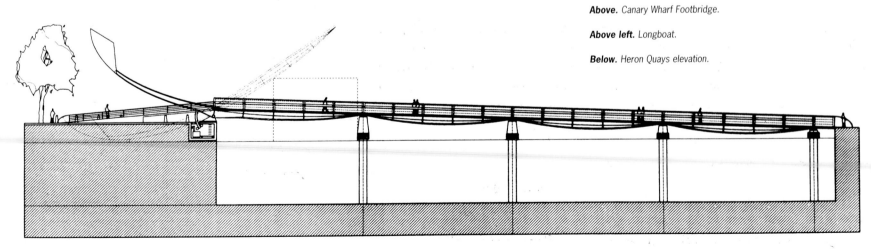

with a counterweight arm almost the length of the opening span, shaped like a tail fin. The counterweight arm and the opening span of the bridge are rigidly connected together to make one long structure. The structure can pivot on a central hinge bearing fixed on the dock wall. To raise the bridge span, an electrically driven screw jack pulls the counterweight arm downwards. The counterweight arm gradually lowers below dock level into a purpose-built chamber, as the bridge span rises.

We engineered the bridge spans to be simply supported so that they could easily be relocated or replaced at some time in the future. The spans can be lifted using recessed built-in lifting points and then moved by barge once the parapets and walkway deck are removed. With careful planning the bridge spans can all be moved and relocated over a weekend.

Above. Heron Quays.

Right. Timber hull construction.

Far right. Steel hull construction.

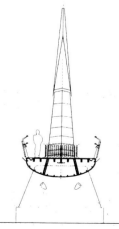
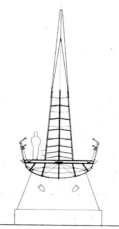

The purpose of this bridge is to transport excavated material from the Deutsche Bundesbahn tunnel to the other side of the River Main. It is also a crossing for the site workers. The design of the bridge was such as to allow it to be easily dismantled and reused, and to minimise any damage to its surroundings. The need for functional economy produced a bridge design without frills.

Nantenbach Bridge Conveying Material and People

Jörg Schlaich

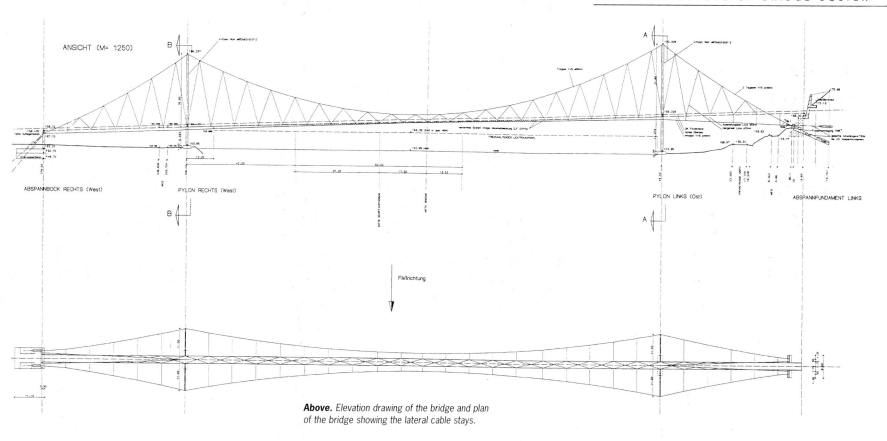

Above. Elevation drawing of the bridge and plan of the bridge showing the lateral cable stays.

The very light longitudinal girder of this bridge has a relatively high wind resistance because of its shape and railings. In order to stabilise the 177m span, the bridge is constructed as a three dimensionally prestressed cable support system. The main central cable, the diagonal hangers and the longitudinal girder make a torsion proof "tube" with a triangular cross section. To secure the bridge against uplift and side winds, the longitudinal girder is braced with two cables arranged laterally and diagonally beneath the girder deck.

Double masted A frame pylons support the main cable and restrain the bridge against sidesway. It is interesting to note that the number of stress changes in this cable-supported system are small. For this reason the diagonal hangers can be anchored to the main central cable using shackles and eye fittings. The cables themselves are anchored to the pylons using cable sockets welded into the pylon tubes.

The longitudinal girder was fabricated in 20m lengths and lifted into position using a cable crane with an additional cable suspended between the pylon tops after the pylons, the main cable and hangers were in position.

Atlanta Footbridge Bridging the Gulch

Jane Wernick, Engineer, Ove Arup and Partners; Jay Thomson, Architect, Thompson Ventulett Stainback

The idea of a suspension bridge that would soar over a huge, grimy, multi-storey car park and the urban gulch below—expressing in a single gesture the natural path between two destinations—was difficult to resist. Much of the appeal lay in the incongruity of placing such a graceful structure in such an industrialised landscape, and in the fact that it does not span a picturesque river but the crude environment of a city car park.

The project began out of a desire to create a pedestrian connection of some sort between two significant areas of downtown Atlanta, separated by a man-made wasteland known as the Gulch. The study was commissioned by the Corporation for Olympic Development in Atlanta, who was responsible for making infrastructure improvements to the city prior to the 1996 Olympics. On one side of the gulch is the plush sports, entertainment and convention complex anchored by the Georgia Dome. On the other side is Underground Atlanta, a district of retail and restaurant developments, targeting visitors to the city. In between, on the site of a former railway goods yard, are elevated roadways, a large parking structure and live railway tracks.

The Omni MARTA rapid rail station which serves the convention centre, the Dome and basketball stadium is presently too small to deal with an Olympic crowd at peak times. Rather than expand the station it was thought prudent to encourage pedestrians to walk a further 1300 ft, over a footbridge or along a ground level walkway, to reach the next station: Five Points. A ground level access would take the pedestrian through a dark and dimly lit car park, and over live rail tracks, which would feel and look unwelcome. Hardly the right image for an Olympic setting.

We naturally preferred the footbridge solution and came up with the idea of a suspension bridge soaring above everything around it. It created a safe haven for the pedestrian, avoiding the jungle of urban excess and dereliction below.

It was also important that we should challenge the conventional form for a suspension bridge and design an innovative structure that would befit a landmark status. The curved alignment of the walkway, the inclined cable masts, and the dynamic twisting of the cable stays of the design seem to accomplish this. Lateral forces set up by the curved bridge deck, were resisted effectively by the inclined cable masts.

The walkway approaches at each end of the bridge are set 20 ft above ground. At the centre of the span the deck is 60 ft above the ground. The walkway exits on to monumental staircases which lead down to the plaza and ground level. The middle span of the walkway curves gently upwards to a point about 20 ft above the car park. The 513 ft length of the middle span offers a view of the skyline and a good sense of where one is going. The cable masts, 140 ft high, provide a focal point for pedestrians traversing the "cat cradle" like alleyway created by the twisting cable stays.

We think this will be the largest suspension bridge over a car park, if and when it is built.

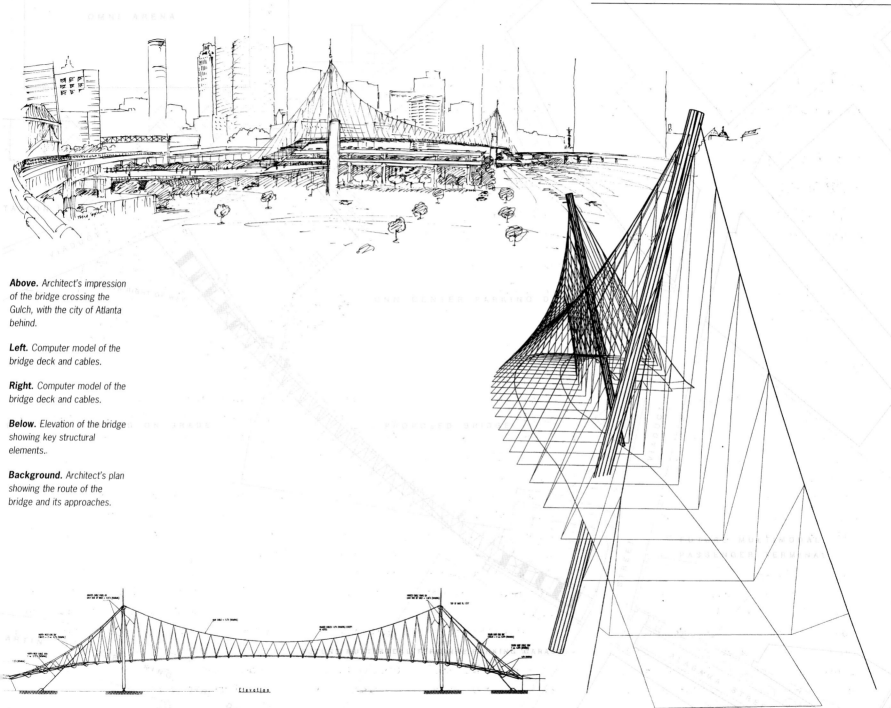

Above. Architect's impression
of the bridge crossing the
Gulch, with the city of Atlanta
behind.

Left. Computer model of the
bridge deck and cables.

Right. Computer model of the
bridge deck and cables.

Below. Elevation of the bridge
showing key structural
elements..

Background. Architect's plan
showing the route of the
bridge and its approaches.

Elevation

Bankside to St Paul's Exploring Ideas for a Crossing

Foster and Partners

This study is an exploratory illustration of ideas, not a definitive design proposal for a new footbridge across the Thames, between St Paul's Walk and Bankside Power Station; the home of the new Tate Gallery. As London's only stand-alone pedestrian and cycle crossing, it should be a special experience.

In developing concepts for the bridge we discussed our ideas with the Port of London Authority, to determine the restrictions for navigation channels on the river. We recognised that the bridge is directly in line with St Peter's Hill steps and the last remaining riverside view of St Paul's. Any ideas that we proposed must be technically feasible and must not impact on the view of the Cathedral.

We configured two solutions that would meet the technical criteria we were set: a bridge with a long central span or one with a series of smaller spans, in keeping with tradition along the Thames. We thought that the large span option would lead to greater support structure cost and a more elevated bridge deck which could impact on the sight line of St Paul's. We therefore focused our ideas on a bridge with three short central spans and two smaller end spans. This is our statement.

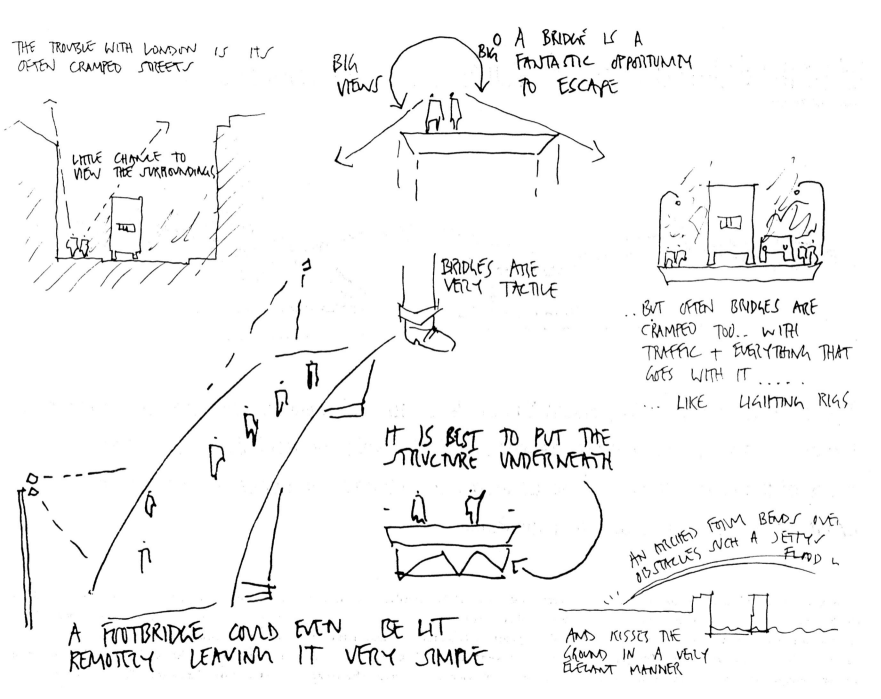

THE TROUBLE WITH LONDON IS ITS
OFTEN CRAMPED STREETS

LITTLE CHANCE TO
VIEW THE SURROUNDINGS

BIG
VIEWS

BIG

A BRIDGE IS A
FANTASTIC OPPORTUNITY
TO ESCAPE

BRIDGES ARE
VERY TACTILE

..BUT OFTEN BRIDGES ARE
CRAMPED TOO.. WITH
TRAFFIC + EVERYTHING THAT
GOES WITH IT .....
... LIKE   LIGHTING RIGS

IT IS BEST TO PUT THE
STRUCTURE UNDERNEATH

AN ARCHED FORM BENDS OVER
OBSTACLES SUCH A JETTY /
FLOOD

A FOOTBRIDGE COULD EVEN BE LIT
REMOTELY LEAVING IT VERY SIMPLE

AND KISSES THE
GROUND IN A VERY
ELEGANT MANNER

A CLEAR SIMPLE DECK....

...LIKE AN UNDULATING CLIFFTOP WALK

CLEAR VIEWS OUT

AN ELEGANT STRUCTURE TO SPREAD OUT THE LOADS

A TRANSFER STRUCTURE AT PIN-JOINT FOUNDATION

MAIN CLEAR SPAN

STREAMLINED BASES TO NAVIGATE THE RIVER

VIEWS TO WIDER LONDON CONTEXT... AND BIG SKIES

A RHYTHMIC PROGRESSION TO ST. PAULS

CLEAR VIEWS TO RIVER + ACTIVITY

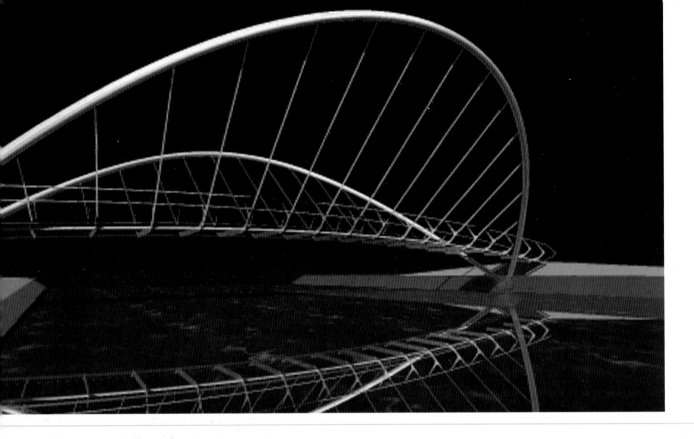

The Embankment Renaissance Footbridge

Keith Brownlie, Architect

*J.J. Webster's 1888 Suspension Bridge provides
a memorable landmark which epitomises
Bedford at the end of the nineteenth century.
The challenge is to create a bridge which
similarly marks the end of the twentieth century
and the beginning of the twenty first.*

Above. *Webster's 1888 Suspension Bridge.*

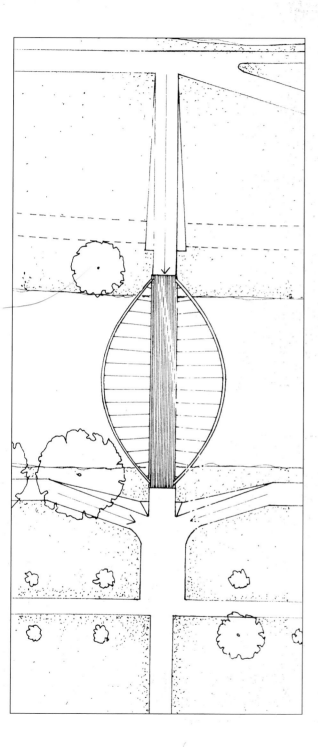

Left. *Plan of the bridge.*

Below. *Bridge in section. The form alludes to the open wings of a butterfly.*

The River Great Ouse is a prime element in the character of Bedford, flowing through the flood meadows which skirt the town as two channels, the Upper and Lower Rivers. The Embankment Renaissance Footbridge is set on the Upper, spanning 30m between willow-lined grassed banks, linking Russell Park to Longholme Island, and adds to a series of existing crossings adjacent to the magnificent Victorian Bedford Suspension Bridge.

The site is inherently calm and rural, being a popular recreational promenade and the subject of extensive preservation works under the banner of "The Embankment Renaissance". An adjacent boathouse animates the river with competitive rowing traffic which characterises this stretch of water and indeed the town, most evidently during the bi-annual River Festival and Regatta.

The design responds to the fundamental objective of the brief: "J.J. Webster's 1888 Suspension Bridge provides a memorable landmark which epitomises Bedford at the end of the nineteenth century. The challenge is to create a bridge which similarly marks the end of the twentieth century and the beginning of the twenty first."

Webster's bridge is a thoroughbred steel structure which

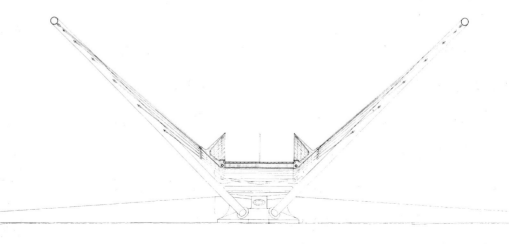

reveals its author's enlightened understanding of materials and structure in an aesthetic framework. Its lightness and presence derive from the paring down of elements in an interactive structural system. All elements function structurally and combine in a cohesive composition which spans with apparent ease and a grace that does not interrupt its context. It will be the job of the new structure to justify itself by reference to this earlier bridge as well as to the characteristics of its wider context.

The Embankment Renaissance Footbridge translates Webster's basic structural system into an evolutionary design applying contemporary engineering and material capabilities. The Suspension Bridge's twin arched trusses are revisited as a pair of high arching parabolas emanating from a single point on each bank, canted to form a splayed opening alluding to the wings of a butterfly. The composition is thus vaguely organic, like some mechanical insect landed on the flood meadows with evident visual reference to its neighbour firmly locating the design in its setting.

Beneath the arches, rods suspend structural "slings" combining deck support members with structural balusters, between which a shallow arched timber deck spans. The deck terminates as it passes through the split arches, adjoining landings cantilevered from expressed concrete footings at each bank.

At the base of the raked balustrade a steel footplate, harbouring fibre-optic deck lighting, encourages spectators to utilise the edges of the bridge to watch the rowers below. The act of crossing is thus enlivened, with the bridge even becoming temporarily formalised as a "grandstand" during river events.

The form invites a number of experiences in its use, being designed for the perception of those passing under as well as over it. The arches dramatically overhang the river and the rowers beneath while the same device opens up to define the route for pedestrians above.

The bridge's purpose, a level and navigable crossing of the Great Ouse, is subsidiary to the more esoteric aspects of its design. Structurally the composition is at the limits of its capability. Visually it is intentionally arresting, and historically it is anchored to its site by reference to the existing.

The design is a result of the collaborative application of art and engineering in contrast to Webster's engineering, latterly perceived as art.

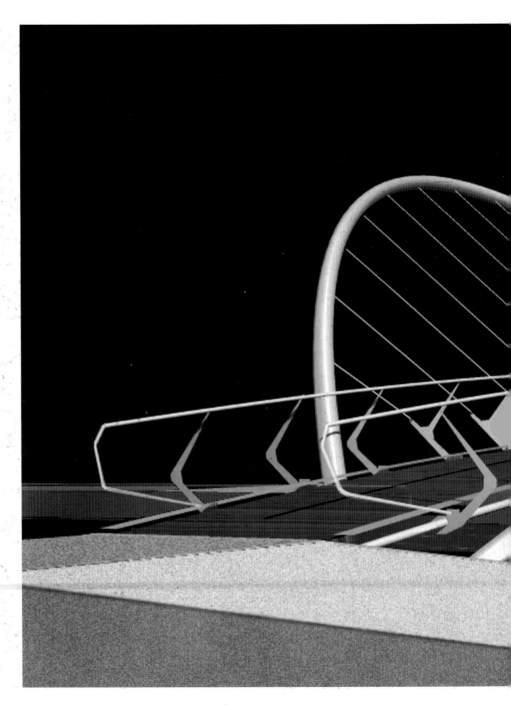

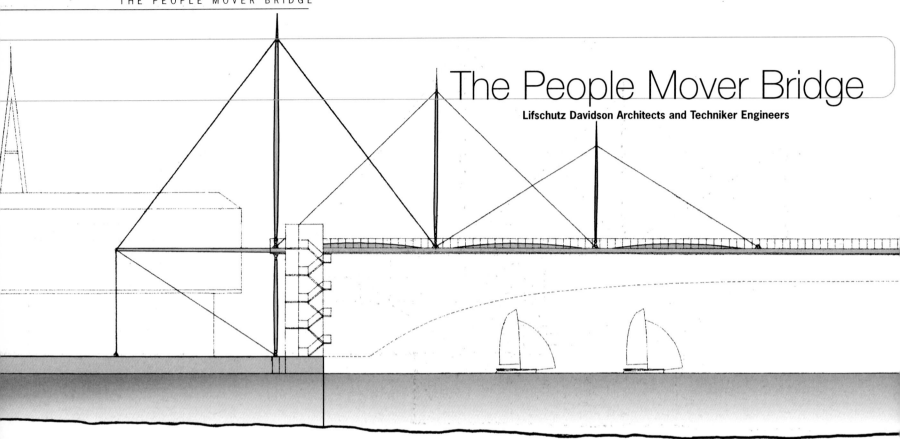

The People Mover Bridge
Lifschutz Davidson Architects and Techniker Engineers

The transporter bridge is not a new idea; it was first suggested by the English engineer Charles Smith in 1873 and later pioneered by the French engineer Arnodin. It was a structure that was subsequently adopted for deep water crossings, where the frequent passage of shipping compromised an opening bridge. We designed our bridge to transport people, rather than vehicles, across a waterway, snug inside a well lit cable car.

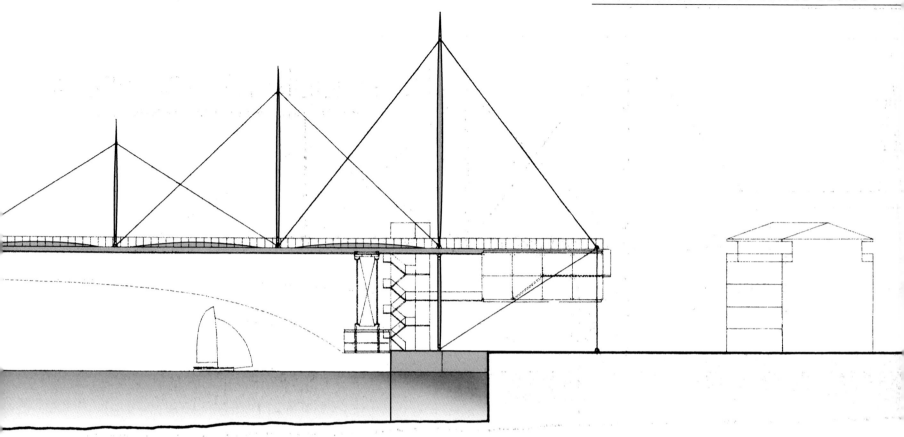

Context and Connections

The design seeks to resolve the inherent conflict of needs between the bridge and the site: on the one hand providing a bridge of 127m span over the docks which meets the brief because it is "a structure of excellence which will contribute to the urban environment" whilst on the other hand it must cause the minimum disturbance to the leisure activities of the waterway below.

We based our ideas on the transporter bridges of a bygone era, drawing inspiration from examples of Arnodin's work at Nantes in 1903 and the cable stayed structures in Marseilles and Brest, designed by Arnodin and Imbault. So the people mover bridge evolved for the Royal Victoria Docks—a lightweight structure with a high level fixed crossing to avoid the protracted opening and closing times of a long span and the complex

engineering of the mechanism needed to operate it. The more bits that are needed to move, the greater the risk of things going wrong. So we kept it simple, moving the people instead of the whole bridge.

The people mover consists of a glazed cabin hung from a moving carriage fixed to the underside of the deck structure. It is suspended so that it can skim across the dock just above water level when there are no boats

in the vicinity. At other times it can be raised to bridge deck level and driven across like a monorail. The people mover always starts and finishes at ground level on reaching the other side. What happens in between is decided by the operator based on wind conditions and proximity of sailing craft.

Pedestrian access onto the footbridge is by lifts or staircases at each end. On race days or special event days, the bridge effectively becomes a viewing platform from which to look out over the boats and waterway. The pedestrian bridge offers a wide walkway with unrestricted views over the docks, but as it is uncovered, it will not be used much in poor weather or on wintry days.

The bridge structure has been designed to relate to a working docklands setting, of gantry cranes, ship masts and cable lines. This is reflected in the tall cable masts above the bridge deck, the box beam lattice structure, and the transport feature of the bridge. The cable mast and lightweight deck capture the spirit of boat building technology, of bean pole masts, guide ropes, winches and timber decking—light and tough, aerodynamic and strong.

Fink Trusses

An arrangement of seven box girder beams supported by cable stays suspended from a series of masts tied back at their ends, form the principal components of the bridge.

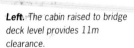

Left. The cabin raised to bridge deck level provides 11m clearance.

Right. The Laser 5000 has a mast height of 8.5m.

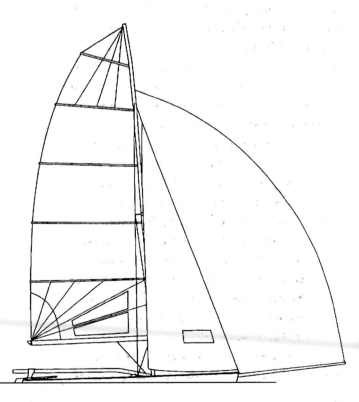

The box girder beams form the main span, supporting one another in succession to make a fink truss. The box girder is a robust and versatile structure much used in marine engineering, ship superstructures and dockyard cranes. In this case it had been modified to a fink truss to perform the composite functions of a column support, a beam and torsion tube.

But why a fink truss? It was a cleverly conceived construction that was designed at the turn of the century to form a lightweight structure out of wrought iron and timber that was strong enough for handling comparatively light loads over relatively long spans. The fink truss behaves like the skeletal frame of a vertebrate animal. The vertebrae and ribs form the compression members, with the spaces between the ribs, criss-crossed by a network of muscles and tendons, the tension components. The fink truss becomes a more rigid structure when it is braced along the bottom chord with a series of tension members. Such structures were used to form the wings of the early strutted biplanes that were flown by the Wright brothers and continued to be built well into the 1930s. Lightness of construction over a long span is why we also chose the fink truss concept.

The lateral stability of the bridge, its resistance to wind loads and the moving load of the cable car, is provided by the box beams acting together as one homogeneous structure; helped by perimeter ties along each side of the deck. The upper shrouds of the main support mast resist the lateral and out of balance loading in the cable stays. A light precompression is applied to the principal cable mast to minimise any elastic shortening of the mast under load.

The longitudinal stability of the bridge to resist the disturbing force due to the acceleration and deceleration

Top. *The musculoskeletal systems of vertebrate animals resemble fink trusses.*

Bottom. *Tension chords improve the performance of fink trusses.*

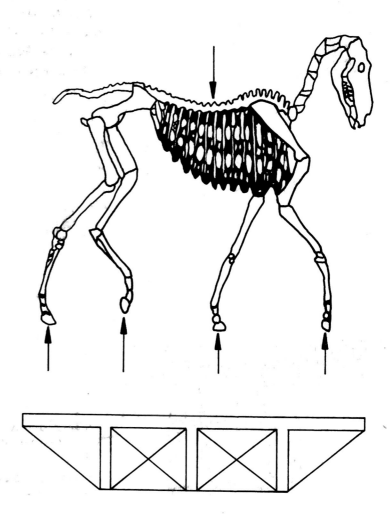

of the cable car, is provided by the braced end bays and tie backs at each end of the bridge. To minimise localised deflections of the bridge under the moving cable car, a prestress to the under-girdle of the box beams has been

applied as well as a mild precamber.

One critical consideration in the design was dealing with the expansion and contraction of the bridge, caused by changes in shade temperature of around 40 degrees.

The generally determinate elements of the structure, the cable mast and stays, will grow larger or smaller proportionally, so that the equilibrium of the internal forces remains stable with no change in the deflected

Below. *A fink truss bridge near Lynchburg, Virginia, USA.*
Copyright: Smithsonian Institution

shape. However the box girder beams forming the bridge deck are restrained against longitudinal movements by the back stays at the ends of the span. Together with the back stays and end ties, these elements are the indeterminate parts of the structure. A big rise in temperature will cause the bridge beams to lengthen cumulatively by as much as 110mm. This amounts to an expansion of 55mm at each end of the bridge. As the bridge beam lengthens, the back stays become strained and tighten up sufficiently to lift the centre spans. By forming the back stays in pre-stretched wire strand, this movement can occur without affecting the upper cable stays that support the box beams. In this way the structure can maintain the required amount of longitudinal rigidity to accommodate this temperature movement.

The main support trestles of the bridge at the two ends, are located within the existing dock and are supported on piled foundations. The ties holding down the back stays are anchored by tension piles formed in the London Clay.

The pedestrian walkway of the bridge is a solid timber decking fixed with self tapping screws to angle frames. The balustrading slots in between brackets welded to the angle frames and then bolted to the main box beam sections.

Distraction for Fun

We have designed the bridge to amplify the notable leisure features that are being planned for this site and environment in the near

future. The bridge is positioned so that it starts at the focal point of the New Urban Village to the south. It crosses the waterway in a diagonal line to meet the corner of the proposed extension to the dock in the north and the pathway leading to the Dockland Light Railway station.

The interactive quality of the bridge, the people mover rides and the spectacular views of the docks, of Greenwich, the Thames Barrier and Canary Wharf Tower should act as a major draw for tourists. Flags and pennants are provided for the masts and cable stays so that the structure can be decked out on race days or special events.

By night, the bridge is lit up to emphasise the height and form of the structure. Downlighters pick out the tall cable masts, while concealed fittings on the balustrading spill light beams through the open timber decking to highlight the whalebone structure of the box beam. The glow of the people mover tracing an arc across the water, forms the backdrop to this dramatic setting.

The bridge provides the developers of the Exhibition Centre and Urban Village in the Royal Victoria Docks with a memorable landmark, and the opportunity to attract substantial numbers of visitors and boat enthusiasts to the area.

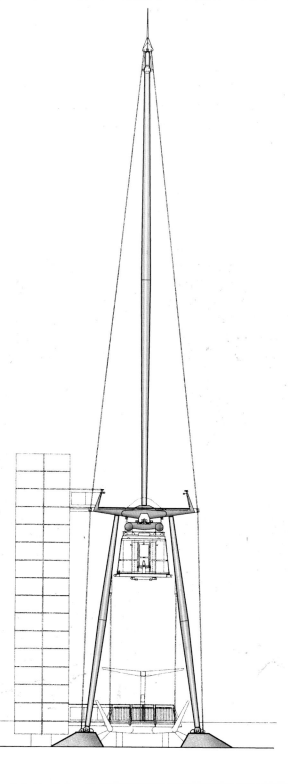

Left. Bridge section.

Technical Data

Bridge Name   Pont Devenir

Status   Competition entry 1994

Location   Geneva, Switzerland

Type of Structure Continuous girder frame of variable inertia, footbridge, road bridge, railbridge

Maximum Span   325m

Overall Length 1600m

Height   35m

Cost   140 million SFR

Design Team   Architects: Rodolphe Luscher; Filippo Broggini; and Pascale Amphoux
Engineer: Jean Tonello

Client   State of Geneva

Bridge Name   Kirchheim Motorway Overbridge

Status   Completed 1993

Type of Structure Prestressed concrete frame

Maximum Span   61.8m

Overall Length 85.0m

Height   9.6m

Cost   1.7 million DM

Design Team   Engineer: Schlaich Bergermann und Partner
Contractor: Richard Besemer

Client   Federal Republic of Germany

How to find the bridge On A8 at Kirchheim near Stuttgart

Bridge Name   Pont Isère

Status   Completed December 1991
Silver medal 1993
FIP award for outstanding structures 1994

Location   Highway A49, France

Type of Structure Cable stayed, road bridge

Maximum Span   148m

Overall Length 304m

Height   Deck 37.5m
Pylon 57.3m

Design Team   Engineer: Jean Muller International and Scetauroute
Architect: Alain Spielmann
Contractor: Bouygues

Client   AREA

How to find the bridge A49 highway Valence-Grenoble, near the town of Bourg-de-Péage

Bridge Name   A75 Clermont Ferrand Viaduct

Status   Completed 1992
Gold medal 1993

Location   A75 Highway near Garabit, France

Type of Structure Concrete road bridge

Maximum Span   144m

Overall Length 308m

Height   80m

Design Team   Engineer: A.I.O.A.; CETE Lyon; SETRA; E.E.G.
Architect: Alain Spielmann
Contractor: Dumez and GTM

Client   Ministry of Transport

How to find the bridge A75 highway, near the town of St Flour, about 2km from Eiffel's Garabit Bridge

PAGE 12

PAGE 16

PAGE 18

PAGE 24

Bridge Name | **Puente de Sancho el Mayor Sobre el Rio Ebro**

Status | Completed 1978

Location | Highway of Navarra, Castejón, Spain

Type of Structure | Cable stayed, box girder, concrete prefabricated, road bridge

Maximum Span | 146m

Overall Length | 203m

Design Team | Engineer: Carlos Fernández Casado, S.L. Contractor: Dragados y Construcciones & Huarte y Compañia

Client | Autopistas de Navarra S.A.

How to find the bridge | Near Castejón Village on the Navarra highway, over the Ebro river

PAGE 36

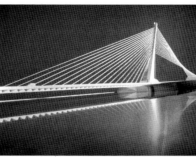

Bridge Name | **Ponte Sobre o Rio Lerez en Pontevedra**

Status | Completed 1995

Location | Pontevedra, Spain

Type of Structure | Cable stayed, box girder, concrete cast in place, road bridge

Maximum Span | 126m

Overall Length | 126m

Design Team | Engineer: Carlos Fernández Casado, S.L. Contractor: Ferrovial & Castro Matelo

Client | Xunta de Galicia

How to find the bridge | In the city of Pontevedra upstream of the Santiago Bridge

PAGE 36

Bridge Name | **Twin Mast Bridge**

Status | Completion 1995–96 Citation 41st Annual Progressive Architecture Awards, 1993

Location | Saint Paul, Minnesota

Type of Structure | High strength steel, stainless steel and reinforced concrete cable stayed bridge

Overall Length | 1200ft

Height | 290ft

Design Team | James Carpenter Luke Lowings Richard Kress Janet Fink Engineer: Toltz, King, Duvall, Anderson and Associates, Saint Paul, Minnesota Consulting Engineer: Jörg Schlaich, Schlaich/Bergermann, Stuttgart, Germany

Client | City of Saint Paul Department of Public Works

PAGE 40

Bridge Name | **Årstaviken Bridge**

Status | Rail bridge Construction to start Spring 1997 Competition winner

Location | Årstaviken, Stockholm, Sweden

Type of Structure | Concrete box girder railbridge

Maximum Span | 78m

Overall Length | 800m

Height | 26m

Design Team | Engineer: Banverket Engineers (Swedish Railways) Architect: Foster and Partners Concept Engineer: Ove Arup International

Client | Banverket (Swedish Railways)

How to find the bridge | Sodramalm, 2km south of Stockholm, over the Straits of Årsta

PAGE 46

Bridge Name **Third Runnymede Crossing**

Status Concept design

Location M25 motorway, Egham, England

Type of Structure Steel arched bridge structure designed as a lattice monocoque frame

Bridge Name **Bataille du Texel**

Status Completed October 1994

Location Quai de Risban 59140 Dunkerque, France

Type of Structure Steel structure, two-lane lifting bridge, mobile without counterweights

Maximum Span 28m

Overall Length 60m

Cost (without tax) 22.7 million FF

Design Team Architect: Pascale Seurin, Paris Engineer: Port autonome de Dunkerque Contractors: Baudin-Chateauneuf; SOGEA, SETIEE Masts: MAG, France

Client Syndicat mixte des transports urbains of the urban community of Dunkerque

How to find the bridge Seaward entrance to Dunkerque Harbour near the Minck's Tower

Bridge Name **St Saviour's Dock Bridge**

Status Completed June 1996

Location St Saviour's Dock, London

Type of Structure Cable stayed, steel and timber footbridge

Maximum Span 9m

Overall Length 29m

Height 8m

Cost £620 000

Design Team Engineer: Whitby and Bird Architect: Nicholas Lacey Contractor: Christiani and Nielsen

Client London Docklands Development Corporation

How to find the bridge From Tower Bridge south side, walk east along the riverside. About 100m past the Design Museum

Bridge Name **Lifting bridge over the Gouwe River**

Status Completed in 1989 Dutch Concrete Prize in 1991 EC Best Concrete Bridge category Award in 1992

Location Railway track Utrecht – Rotterdam near Gouda, Netherlands

Type of Structure Railway bridge 2 x 2 tracks (2 tracks/bridge) Bridge 2 lattice moveable bridges Lifting portal concrete/steel

Maximum Span 44m

Overall Length Both sides 300m landbridges

Height 45m

Cost 35 million Nlg including lifting bridge and land bridges

Design Team Engineer: Holland Railconsult, Utrecht, Netherlands Architect: Holland Railconsult, Utrecht, Netherlands Contractor: Dirk Verstoep B.V. Rotterdam, Netherlands

Client N.V. Nederlandse Spoorwegen (Netherlands Railways)

How to find the bridge The bridge is situated just outside the Gouda railway station in the direction of Rotterdam (west of Gouda central station) by the river Gouwe (15 min. walk from the railway station)

THE ARCHITECTURE OF BRIDGE DESIGN

Bridge Name	**Lifting bridge over the Oude Maas**
Status	Completed in 1993 European Steel award in 1995
Location	Railway track Rotterdam – Dordrecht near Dordrecht, Netherlands
Type of Structure	Railway bridge 2 x 2 tracks (2 tracks/bridge) Two lattice moveable bridges and, leading to the moveable bridge, two lattice fixed bridges Steel lifting portal
Maximum Span	88m
Overall Length	2 x 88m
Height	74m
Cost	68 million Nlg including the concrete works, the lifting bridge and one of the lattice bridges of 176m
Design Team	Engineer: Holland Railconsult, Utrecht, Netherlands Architect: Holland Railconsult, Utrecht, Netherlands Contractor: V. Buyck
Client	N.V. Nederlandse Spoorwegen
How to find the bridge	The bridge is situated just outside the Dordrecht railway station in the direction of Rotterdam by the river Oude Maas (10 min. walk from the railway station)

PAGE 69

Bridge Name	**South Dock Footbridge**
Status	Competition winner Completion April 1997
Location	Spanning South Dock, Heron Quays, London E14, England
Type of Structure	Cable stayed, steel footbridge
Maximum Span	90m
Overall Length	180m
Height	32m
Cost	£2.5 million
Design Team	Engineer: Jan Bobrowski Architect: Chris Wilkinson Architects Contractor: Christiani and Nielsen
Client	London Docklands Development Corporation
How to find the bridge	Heron Quays DLR, Canary Wharf JLE (when completed)

PAGE 74

Bridge Name	**Lowry Centre Lifting Footbridge**
Status	Competition shortlist concept design July 1995
Location	Manchester, England
Type of Structure	Steel footbridge
Maximum Span	90m
Height	6m (lifted 23.5m)
Overall Length	110m
Design Team	Engineer: D.H.V. Architect: Alain Spielmann
Client	City of Salford

PAGE 80

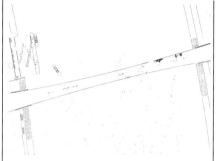

Bridge Name	**The Great Arch over The Danube**
Status	Competition entry 1994 1st prize 1994 Concept design
Location	Budapest, Danube crossing
Type of Structure	Metal lattice structure, footbridge
Maximum Span	385m
Overall Length	500m
Height	35m
Design Team	Architects: Rodolphe Luscher; Filippo Broggini; and Andreas Lorenz Engineers: CET Budapest
Client	City of Budapest

PAGE 84

Bridge Name	**Lintas Bridge**
Status	Completed 1983
Location	Quai de la Magister, Paris
Type of Structure	Cable stayed, steel and glass
Maximum Span	20m
Overall Length	20m
Height	1m (on 4th floor)
Cost	£35 000
Design Team	Engineer: RFR Architect: Ian Ritchie
Client	Lintas
How to find the bridge	Via Lintas offices — on 4th floor

Bridge name	**Ecology Gallery Bridge**
Status	Completed 1991
Location	Natural History Museum, London
Type of Structure	Cable truss, steel, timber and glass. Exhibition/pedestrian bridges
Spans	3 x 6m 1 x 9m
Height	3m
Design Team	Engineer: Arup Architect: Ian Ritchie Architects Contractor: Thanet Foundry & Engineering Co Ltd
Client	Natural History Museum
How to find the bridge	Natural History Museum Ecology Gallery

Bridge Name	**Science Museum Bridge**
Status	Started on site December 1996 To complete April 1997 Competition winner
Location	Science Museum, London
Type of Structure	Footbridge
Span	14m
Overall Length	14m
Cost	£200 000
Design Team	Engineer: Whitby and Bird Architect: Chris Wilkinson Architects
Client	Science Museum
How to find the bridge	1st floor, Challenge of Materials Gallery, Science Museum, London

Bridge Name	**Earth Bridge, Croydon (The Garden Bridge)**
Status	Concept bridge developed in 1992, not being constructed
Location	Croydon, pedestrian crossing above Park Lane linking the Queens Gardens and the Fairfield Halls
Type of Structure	Two arches constructed in polymer stabilised top soil or any other powdered matter (Jack Blackburn Patent)
Spans	2 x 22m
Overall Length	87m
Height	9m maximum height (excluding trees)
Design Team	Architect: Cezary M Bednarski Engineer: Tim Macfarlane/ Dewhurst Macfarlane and Partners Advisor: Jack Blackburn
Client	Proposal to Croydon Council

PAGE 87

PAGE 90

PAGE 94

PAGE 98

Bridge Name	**Beaune Bridge**	**Bridge Name**	**Le Grand Viaduc de Millau**
Status	Completed April 1991	**Status**	Concept design stage Competition winner Will be under construction before 2000
Location	A6 highway, town of Beaune, (Burgundy Region)	**Location**	Gorges du Tarn, Millau, Aveyron, France
Type of Structure	Cable stayed road bridge	**Type of Structure**	Concrete cable stayed road bridge
Maximum Span	34.71m	**Maximum Span**	350m
Height	(pylon) 25m	**Overall Length**	2500m
Overall Length	64.96m	**Height**	275m (to deck)
Design Team	Engineer: Michel Placidi (Razel) Architect: Alain Spielmann Contractor: Razel	**Cost**	1.4 billion FF
Client	S.A.P.R.R.	**Design Team**	Engineer: Sogelerg, EEG and Serf Architect: Foster and Partners with Chapelot Monssegene Defol
How to find the bridge	On the A6 highway, town of Beaune, in the heart of the Burgundy wine-producing area	**Client**	Arrondissemont Interdepartemental des Ouvrages D'Art
		How to find the bridge	Along the A75 Clermont Ferrand highway near Millau, over the River Tarn in the Aveyron Region of France

Bridge Name	**A14 Nanterre Viaduct**
Status	Motorway viaduct Completed 1995 (viaduct), 1996 (building)
Location	Nanterre, Hauts-de-Seine, France
Type of Structure	Prestressed concrete deck on steel piers
Maximum Span	35m
Overall Length	250m
Height	10m
Design Team	Engineers: RFR with Ove Arup & Partners (viaduct), Terrell Rooke Associés (building) Architect: Odile Decq & Benoît Cornette Contractors: ETPO/El/Levaux (concrete), Victor Buyck (structural steelwork)
Client	Etablissement Public pour l'Aménagement de La Défense (EPAD)
How to find the bridge	From Paris, take RER (suburban express railway) line A, towards St Germain en Laye, to Nanterre Prefecture. From the RER station walk westwards towards the Seine. For a view of the bridge from above, stay on the train for another stop.

Bridge Name	**A13/A406 Gateway Structure**
Status	Concept proposal under consideration
Location	Beckton, East London, England
Type of Structure	Cable stayed road bridge
Maximum Span	90m
Overall Length	253m
Height	(pylon) 72m
Cost	£8.9 million
Design Team	Engineer: Babtie Group Consulting Engineers Architects: Nicoll Russell Studios
Client	London Docklands Development Corporation
How to find the bridge	Junction of the A13 and A406 in Beckton

PAGE 102
PAGE 118
PAGE 134
PAGE 138

Bridge Name	"Islands in the water" (Canary Wharf/Heron Quays footbridge)
Status	Competition entry Concept design
Location	Canary Wharf/Heron Quays, Isle of Dogs, London
Type of Structure	Pontoon, concrete with steel and timber footbridge
Maximum Span	11.4m
Overall Length	140m + 57m
Height	3m (8m raised)
Cost	£1.7 million
Design Team	Engineer: Ove Arup & Partners Architect: Ian Ritchie Architects
Client	London Docklands Development Corporation

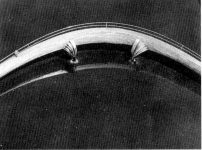

Bridge Name	East Footbridge, La Villette
Status	Public footbridge Completed 1992
Location	Parc de la Villette, Paris, France
Type of Structure	Steel truss footbridge
Maximum Span	45m
Overall Length	100m
Height	8m
Cost	6 million FF
Design Team	Engineer: RFR Architect: Bernard Tschumi Contractor: Cegelec
Client	Etablissement Public du Parc de la Villette
How to find the bridge	The bridge crosses the Canal de l'Ourcq in the public park at La Villette in the 19e district of Paris. Métro stations: Porte de la Villette or Porte de Pantin.

Bridge Name	Leith Footbridge
Status	1994 Winner—Swedish Timber Competition (for innovation in timber design)
Type of Structure	Timber footbridge
Maximum Span	15m
Overall Length	50m
Design Team	Architect: Clash Associates Engineer: Techniker

Bridge Name	Landmark Bridge for the Albert Dock, Liverpool
Status	Study analysis 1995/96
Location	The Albert Dock, Liverpool, linking Edward and Atlantic Pavilions
Type of Structure	A semi-monocoque aluminium beam composed of four toroidal surfaces off which a perforated steel trough carrying shot blasted glass walkway is suspended
Span	40m
Overall Length	40m, 7.5m height above dockside
Design Team	Architect: Cezary M Bednarski Engineer: Matthew Wells/Techniker Artist: Peter Fink
Client	Arrowcroft

PAGE 146

PAGE 152

PAGE 154

PAGE 158

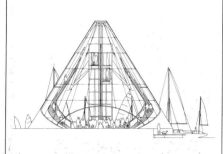

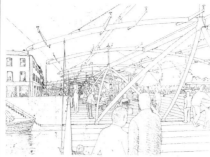

Bridge Name	**The Royal Victoria Dock Pedestrian Crossing, London**
Status	Finalist in a limited competition entry
Location	The Royal Victoria Dock, London
Type of Structure	Fully weather protected bridge comprising two glazed steel plate cantilevers, both ends of the bridge in form of glazed semi-cons housing stairs, lifts and cantilever tendons
Span	150m
Overall Length	150m, 13m height clearance for sailing, 18m height above dockside
Design Team	Architect: Cezary M Bednarski/ Studio E Architects Engineers: Dr William Brown/ Brown Beech Tim Macfarlane/Dewhurst Macfarlane and Partners Dr Stafford Craig/Brown Beech Artist: Peter Fink Costings: DLE
Client	London Docklands Development Corporation

PAGE 158

Bridge Name	**Cornmarket Canopy Bridge**
Status	Possible construction 1998
Location	Cornmarket, Cork, Eire
Type of Structure	Steel truss, covered, footbridge
Span	40m
Overall Length	40m
Height	4m
Cost	£400 000
Design Team	Engineer: Ove Arup & Partners Architect: Ian Ritchie Architects (Design) Others: Executive architects: Kelly Barry O'Brien QS: P. J. Coveney Partnership
Client	Cork Corporation
How to find the bridge	(When built) North end of Cornmarket, Cork City Centre (northern branch of River Lee)

PAGE 162

Bridge Name	**Japan Bridge**
Status	Public footbridge Completed 1993
Location	La Défense, Paris, France
Type of Structure	Steel tied arch footbridge
Span	100m
Overall Length	100m
Height	15m
Design Team	Engineers: RFR with Ove Arup & Partners Architect: Kisho Kurokawa Contractor: Viry
Client	SARI Construction
How to find the bridge	From Paris, take RER (suburban express railway) line A, towards St Germain en Laye, to La Défense, or Métro line 1 to Grande Arche de La Défense. The bridge is around 200m beyond the Grande Arche (to the west); walk around the arch to its south side, following signs to the Tour Pacifique.

PAGE 164

Bridge Name	**Canary Wharf Footbridge**
Status	Competition entry
Location	Canary Wharf, Heron Quays, London
Type of Structure	Steel/timber footbridge
Maximum Span	20m
Overall Length	90m
Cost	£1.5–2 million
Design Team	Engineer: W S Atkins Consultants Ltd Architect: Nicholas Grimshaw and Partners Other consultants: Hayes Davidson (computer rendering)
Client	London Docklands Development Corporation

PAGE 168

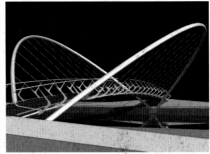
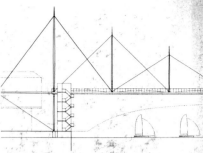

Bridge Name	**Nantenbach Bridge**
Status	Temporary structure 1990
Location	Nantenbach, Germany
Type of Structure	Cable suspension bridge
Maximum Span	177m
Overall Length	279m
Height	39m
Design Team	Engineer: Schlaich Bergermann und Partner Contractor: Pfeifer Seil und Hebetechnik Others: Stuag, Züblin, WTB
Client	Arge Schönraintunnel
How to find the bridge	Dismantled

Bridge Name	**Atlanta Footbridge**
Status	Pedestrian bridge, concept
Location	Between the Omni Plaza and Alabama Street, Atlanta, Georgia, USA
Type of Structure	Suspension bridge, steel masts and cables, concrete deck
Spans	513 ft
Cost	US$5.5 million, approx.
Design Team	Engineer: Ove Arup & Partners Architect: Thompson Ventulett Stainback
Client	Committee for Olympic Development in Atlanta

Bridge Name	**Embankment Renaissance Footbridge**
Status	To be completed October 1997 Competition winner
Location	River Great Ouse, Bedford, England
Type of Structure	Cable stayed footbridge
Span	28m
Overall Length	28m
Height	7.5m
Cost	£380 000
Design Team	Engineer: Jan Bobrowski & Partners Architect: Chris Wilkinson Architects
Client	Bedford Borough Council
How to find the bridge	East of town centre on River Great Ouse

Bridge Name	**Royal Victoria Dock Footbridge**
Status	Competition winner September 1995 Due for completion December 1997
Location	Royal Victoria Dock, North Woolwich Road, London
Type of Structure	Cable stayed steel footbridge with people transporter
Maximum Span	127.5m
Overall Length	178.5m
Height	15m to bridge deck 45m to top of main mast
Cost	£3.8 million
Design Team	Architect: Lifschutz Davidson Structural Engineers: Techniker M & E/Transporter Design: Allott & Lomax
Client	London Docklands Development Corporation
How to find the bridge	Public Transport—Take the Docklands Light Railway to Custom House Station By car—Take A13 and turn onto Silvertown Way at Canning Town

The Lerez bridge.